Fionn Stevenson

C000097067

[Handwritten notes:]
EVERY HOUSE NEEDS.
FEEDBACK / TESTING / BPE
· FIND CASE STUDY eg ARCHITYPE

AWARE OF THE REALITIES OF THE SITE
EG NOISY ROAD = DIF. VENTILATION + HEATING P61

HOUSING FIT FOR PURPOSE

PERFORMANCE, FEEDBACK AND LEARNING

RIBA ⊞ Publishing

Published by RIBA Publishing, 66 Portland Place, London, W1B 1NT

ISBN 9781 85946 824 1

British Library Cataloguing-in-Publication Data
A catalogue record for this book is available from the British Library.

Commissioning Editor: Elizabeth Webster
Assistant Editor: Clare Holloway
Production: Richard Blackburn
Typeset by: BORN
Printed and bound by Short Run Press Limited, Exeter.

www.ribapublishing.com

Contents

About the Author

Fionn Stevenson holds a Chair in Sustainable Design at the University of Sheffield School of Architecture. She has previously held academic positions in five other UK Universities and was in practice for eight years as a chartered architect. Her research over the last 20 years has focused on developing innovative methods of building performance evaluation (BPE) in relation to occupancy feedback in order to improve building design. She has held a visiting professorship at the University of British Columbia in Canada, and has worked on BPE projects in the UK, Poland, Mexico and Brazil. She is particularly interested in the control interfaces between housing and people from a holistic perspective that includes resource use in its widest dimension. She is also co-director of the Royal Academy of Engineering's Centre for Excellence in Sustainable Building Design in Sheffield. She has obtained and managed £1.53 million in research funding to date, primarily working with UK government agencies and the EU. She has 116 publications to date including books, journal papers, edited special issues, book chapters, and reports. She also serves as a RIBA national role model, encouraging diversity within the profession.

Acknowledgements

As with all books, this one rests on the shoulders of those who have gone before me in developing building performance evaluation in relation to sustainable housing, as well as those who have helped to bring this book into being. You are many, and my thanks to you if you are not named either in the book or below.

A special thanks goes to the following:

My partner, **Margaret Boudra**, you have been there for me every day and without you, it simply would not have happened. Thank you for all your support and many kindnesses.
Richard Lorch, for your thoughtful encouragement and feedback on the entire manuscript.
Howard Liddell (deceased) and **Sandy Halliday**, for sheer inspiration by example and for always pushing my boundaries.
Bill Bordass and **Adrian Leaman,** as founders of the Useable Buildings Trust, you took me under your BPE wings, showed me how to fly, and helped this book come into being.
Julia Park and **Clare Murray,** for their careful reading of the manuscript and insightful suggestions.
Ben Derbyshire, for kindly agreeing to write the Foreword to this book.
Liz Webster and **Clare Holloway** of RIBA Publishing, for coping with my sheer ineptitude with software, and for excellent support and editorial advice.
Richard Blackburn, Head of Production at RIBA Publishing, for nursing the book and me through the final stages of delivery.
All of those interviewed and other contributors via many projects referenced **architects, engineers, physicists, educators, developers**, and those who live with the consequences of housing design, construction and management – the **inhabitants**.

Foreword

It might be expected that the vital priority for investment in good quality, affordable housing would be accompanied by a thorough and systematic approach to obtaining reliable quality outcomes – not so. The production of new, and upgrading of existing housing, generally takes place without widely adopted systems for establishing the performance required, in terms of standard-setting, testing or feedback in use. The consumer is probably better informed about the contents of their daily breakfast cereal than they are about their home. Housing also has more environmental impact overall than any other building type, and comes with significant responsibilities. The dearth of feedback in housing on which to build a convincing body of knowledge and understanding has left the homebuilding industry lagging far behind.

Our fragmented building industry needs to collaborate to produce a holistic response to the issue of housing improvement through learning and feedback. Fionn's book comprehensively shows how to piece together the performance evaluation jigsaw to provide a clear picture of where we need to be. The stand-alone primer on housing performance evaluation is a unique resource that every design team should have at its fingertips.

Fionn sets out clearly why the design, construction and management of housing needs to be linked to feedback on performance in use. Crucially, she includes actual evidence of the way occupants choose their homes, and then learn to live in them. Fionn's approach demonstrates how this can inform the procurement of housing fit for purpose.

For the first time, a sound theoretical basis is provided for housing occupancy feedback, grounded in practical examples. New feedback methods and techniques, as developed by the author, are introduced which relate directly to adapting to and mitigating climate change. Fionn also breaks ground by setting out the training and education needed for delivering effective housing occupancy feedback, and how to embed this in practice. At the same time, she challenges the orthodoxy of current performance evaluation by re-framing how we think about people living in their homes. She challenges designers to work *with* inhabitants, rather than for them.

This book is packed with innovative thinking that takes the reader on an intimate journey into previously unpublished territory concerning the costs and benefits of housing performance evaluation, how to develop in-house practice performance feedback and learning, and how to win over clients and gain repeat design work through embedding performance feedback. It asks difficult and urgent questions in relation to the ethics and risks of obtaining performance feedback, which are answered with guidance based on the author's years of experience, working successfully with house builders and design practices, big and small.

The RIBA has now recommended that all its member practices offer to carry out post-occupancy evaluation on their projects after 2020. This book is a core reference for that process. If we are to transition successfully to an industry geared to the reliable delivery of predictable built performance in terms that are meaningful to consumers as well as professionals, we need tools that are accessible and understandable from all points of view. *Housing Fit for Purpose* takes us significantly closer towards achieving the universal systems for standard setting, predicting, measuring, feedback and learning from outcomes that are needed to tackle the shortfalls in housing performance, promote good design, create human well-being and address the existential threat of irreversible climate change.

Ben Derbyshire, President of the RIBA (2017–2019) and Chair, HTA Design LLP

Introduction

Why should I do building performance evaluation?

Ann Shebold has just returned from an inspiring tour of a home on a housing development for which she is the project architect. This is part of the routine building performance evaluation (BPE) that her practice undertakes on all its projects. The independent BPE evaluator, design team and client representatives met with the inhabitants and discovered that they were overjoyed with the spatial quality of their new home, and loved the views. They particularly liked the natural lighting, and the shading facility on their windows. Ann noticed, however, that the grandmother was struggling to open the window over the sink, and made a mental note to redesign this area more innovatively next time round, and check the handle specification too. She is delighted, however, that the visit has corroborated the 90% satisfaction rating from the inhabitants' questionnaire, which puts the project in the top 10% in the UK database – great for the practice's reputation. She is also relieved that they managed to catch the missing insulation on the eaves of every dwelling, and can have the contractor fix this prior to handover, thanks to the quick thermographic survey undertaken as part of the BPE process. This should prevent thousands of pounds' worth of damage through mould arising from cold bridging, and also save the client's reputation by revealing a hidden problem that would have caused trouble later on.

Back in the office, Ann and the rest of the design team have a debriefing with the independent BPE evaluator on the comparison between the design drawings and what has actually been built, and the other routine and relatively quick 'light-touch' BPE processes that they always go through. They want to know if they have any issues that warrant further diagnostic or even forensic investigations, or if they can just take the valuable design/construction/inhabitation lessons learned so far and move on to the next project. The evaluator says that in this instance, although the homes are generally performing well, the energy usage is significantly higher than anticipated, even though the cost is still relatively low, and that it might be worth drilling down to check what was happening. A further diagnostic investigation reveals that the mechanical ventilation system has been wrongly calibrated, and most inhabitants have simply switched it off. They are quite happy to open the windows instead in the winter to get some fresh air, but this wastes quite a lot of energy. The problem is quickly fixed by recalibrating the systems, and by the client improving their ventilation guidance. The client is delighted when energy levels improve the following year – a chance to go for another award – and, given the due diligence of the architecture practice, a repeat commission is now certain.

The people in this story are entirely fictitious. The story of the poorly designed window, missing insulation and malfunctioning ventilation system is true, however. Fortunately, this fictional developer had an efficient feedback system in place via the routine use of BPE by the architecture practice, and new solutions were found. In reality, very few building owners are able to learn and innovate from building performance in this way – they simply do not have the feedback systems set up to do this (see Figure 0.1) – and as a result, many buildings today continue to drastically underperform, right from the start.

Why has this book been written?

The built environment is responsible for nearly half of all carbon emissions in the developed world, and plays a key role in climate mitigation. New housing, however, routinely uses up to

Figure 0.1 Many are still flying blind when it comes to understanding the performance of their homes

three times more energy than predicted – and this is disastrous in terms of trying to reduce energy use to a level that will lower carbon emissions sufficiently to avoid drastic climate change (see Figure 0.2). There is an urgent requirement to close this yawning 'credibility gap' between predicted and actual energy performance. Housing performance evaluation and feedback need to increase rapidly and provide a comprehensive evidence base to help improve the situation.

But improving housing performance cannot be tackled from a purely scientific perspective. As Janda[1] has eloquently pointed out: people use energy, not buildings. Housing performance evaluation needs to be based on a much broader series of considerations than previous literature in this field has acknowledged. It needs to engage with a wide variety of disciplines within cultural studies, social science and medicine in addition to the usual building science domain. The first half of this book aims to address this broader theoretical underpinning in relation to knowledge exchange between designers, clients and inhabitants, as a means of improving design and housing management, and the learning cycle involved in doing this. The second half provides practitioners with the knowledge to carry out effective evaluations and make these routine.

A key aspect of housing performance evaluation, learning and innovation is the role that time plays. Understanding how homes really work takes time. It is an iterative process which involves revisiting the same building performance problems from different perspectives in order to understand the 'difficult whole'.[2] The author has carried out longitudinal studies that revisit the same housing after a number of years in order to understand how they perform in reality.

In many respects, housing performance evaluation is similar to the housing design process that architects go through, requiring an understanding of different factors. However, whereas design requires a synthesis of these factors in order to produce a coherent whole, performance evaluation requires an analysis of these factors in relation to each other to understand what exactly is going on in the home, and how to improve on this. This process uncovers the reasons why issues are occurring, by 'peeling back the various layers' involved (see Figure 0.3). Looking at the same issue from a number of different angles can build up a rich picture which then reveals 'emergent' problems, hitherto unforeseen but clearly contributing to the performance issues at hand. The issue of 'emergence' in housing performance evaluation deserves greater coverage, as it is often at the heart of building failure.

Figure 0.2 New housing in the UK routinely uses three times more energy than predicted

The 'performance gap' restated

Being an architect trained to develop client briefs as well as design solutions for buildings has given me the ability to interrogate functionality from a social and a physical perspective. I am interested in the interrelationship between people and their physical surroundings beyond functionality, thinking in terms of the *meanings* and *values* attributed to a building and its component parts by various parties. These two factors fundamentally define how we existentially perceive and engage with housing. This brings into play the role of ethics when determining and learning about housing performance, not just in terms of what is examined, but how it is examined and what the predefined limits to the investigation are. As Till points out, architecture is a highly contingent activity, which depends on the cooperation of many others in order to be brought into existence.[3] Housing performance studies also depend on people being willing to have their homes examined. Until now, there has been relatively little discussion of the ethics involved in housing performance evaluation and the sensitivities around this. I will address this in some detail.

Figure 0.3 Evaluating housing performance is a forensic process – peeling back the layers involved

So what's new?

This book presents a new and coherent theoretical approach to housing performance evaluation based on my extensive work in this area over two decades. It builds on the BPE case study approach first developed by Bill Bordass, Adrian Leaman and others in the 1990s, which develops a 'rich context' understanding based on action research. Much has changed since this approach was developed, with a rapid expansion in interdisciplinary studies related to the built environment. This broader interdisciplinary approach significantly enhances housing performance evaluation, particularly in terms of interrelationships, meanings and values. References in addition to building science include industrial ethnography, human anthropology, behaviour psychology, ergonomics, usability engineering, motivation, perception, grounded research, thermal comfort, sustainable design and development, knowledge management and information theories.

For the first time, housing occupancy feedback and learning is presented here as a rapidly developing area, and includes state-of-the-art research that provides new ways of thinking about inhabitants and housing performance. Until now, housing has not been tackled extensively in BPE studies because of the challenges in dealing with this particular typology. Given that housing is responsible for around 15% of all UK carbon emissions, it is clearly a priority area. My work has helped to develop a variety of housing BPE national programmes, and I have also evaluated innovative housing prototypes as well as new building products used within housing. This has resulted in a number of new BPE methods for these areas.

Recent research studies have modelled future climate change and developed strategies for mitigation and adaptation in the built environment. These utilise passive means, where possible, to reduce the impact on buildings and people of a hotter and stormier climate. It is no longer adequate to carry out housing performance studies that use current climate data for evaluation. There is now a pressing need to provide performance evaluation methods which take account of climate change and are able to build in resilience as part of their outcomes. This is a completely new area for housing performance evaluation in practice.

The need for built environment professionals and students to receive training and education in building performance evaluation is addressed in depth here for the first time, drawing on my experience of teaching and mentoring occupancy feedback approaches. This includes developing BPE knowledge transfer in architectural practice, and unique BPE tools to facilitate rapid learning for design iteration purposes. In effect, the use of BPE in this context becomes the driver for continuing professional development.

Bringing together these different theories and approaches, this book aims to establish a new discourse that broadens understanding beyond the current confines of BPE practice. It explores an innovative methodology for housing occupancy feedback developed by the author, and demonstrates its use in practice.

Finally, there is a concise, illustrated and standalone quick primer on how to undertake housing performance evaluation, for the busy practitioner. This is the first time such a primer has been published as far as the author is aware.

Who is this book for?

This book will be of interest to environmental consultants, designers, engineers, surveyors, developers, facilities and construction managers, building developers and clients, policymakers, academics and students, researchers and inhabitants.

The text introduces a new way of understanding BPE for anyone intending to carry out housing studies. It has two purposes: first, as a source of new ideas and insights which can be reflected on and used to develop practice for those who are already familiar with BPE; second, as a practical illustrated handbook which provides a methodology and tools for putting these ideas immediately into practice for those who are relatively new to this territory. This book will help *all* architects and engineers to understand how their design intentions measure up against actual housing performance. It provides a set of tools for investigating projects and improving the next design as a result.

This book will be of particular interest to housing surveyors and managers, as it gives them a means of ensuring that any tacit or emergent problems in their housing portfolio are dealt with systematically using an embedded BPE strategy. This will save time and reduce stress for all concerned. For housing investors, clients and developers, learning about BPE will help them to save cost and reduce risk and liability, effectively futureproofing their investment. It will also provide them with fundamental evidence about the performance of their projects from the outset, which can greatly help them to innovate and improve their housing process and products. For construction and project managers seeking relatively trouble-free building and commissioning processes, key areas are highlighted for project improvement which relate directly to BPE feedback processes. Lessons from practice are related through various case studies.

The hope is that policymakers will read this book as a road map for how government agencies can develop appropriate housing regulatory frameworks with BPE at their core. Without this, it is hard to see how suitable evidence will be gathered to underpin future policies related to housing performance. The small number of housing BPE studies carried out over the past couple of decades in relation to the vast number of housing developments undertaken shows that a voluntary system for BPE does not work, and regulation is needed. Planners, building officers, conservationists, local authority officers and others in government can draw on the theories and tools presented here to develop policies which inherently test and improve local housing stock. Non-governmental organisations, including charities concerned with housing, will also find this book useful to promote their own causes, using the BPE tools and case studies presented here, to help them build their own evidence base.

Academics, researchers and students will find a wealth of material on the performance of newbuild and retrofit housing within the various case studies. Many of the ideas and techniques presented here can be readily deployed in wider housing research projects, not necessarily directly related to BPE. Teachers in higher education institutions can draw on this material to prepare their own modules and programmes for BPE as part of curricula for built environment studies.

And finally, for inhabitants everywhere, I hope this book will provide you with the ammunition you need to check up on the health of your own housing and continuously improve your living and working environments.

How can I use this book?

The book has six discrete but interrelated sections:

1. Background
2. Learning from feedback
3. Training for feedback
4. Application and case studies
5. Challenges for the future
6. Housing feedback primer

These form an intellectual 'journey' through the territory of BPE for the reader, starting out with origin of BPE (Background), gathering the maps needed and packing our belongings (Learning), surveying the terrain (Training) and crossing it (Application), before looking to the next horizon (Challenges). The preparation is set out in the first half of the book, while the second half deals with the actual journey – the practice of BPE. For those less interested in the background and theory, I suggest you move straight to section three onwards. The short, standalone primer at the end of the book describes how to carry out BPE housing studies using the methodology set out in the book, providing a readymade BPE toolkit for the busy practitioner.

The book content is as follows:

1. *Background*

   The first chapter in this section charts a brief history of housing performance evaluation through the international development of this field over the last century, from its first outing in Victorian public health work, right up to the latest interdisciplinary ventures. Various schools of thought are discussed and seminal texts are highlighted. The second chapter provides the context for the book, defining key drivers for housing performance evaluation, including climate change, resource use, legislation, and key initiatives in the UK and Europe.

2. *Learning from feedback*

   This section develops a particular theoretical approach to housing performance evaluation and learning. Chapter 3 sets out BPE building science with specific reference to energy use, thermal comfort and health. It introduces the role of BPE in evaluating housing adaptation for climate change. Chapter 4 uses an interdisciplinary approach to housing performance evaluation to highlight new opportunities and the limitations of current BPE. This connects needs, behaviour, motivation, perception and participatory learning to BPE, drawing on a variety of socio-cultural methods. These are fundamentally grounded in a 'real world' action research approach which involves heuristic enquiry. Chapter 5 describes how this approach can be used to feed back into the housing design and construction process and improve the accuracy of modelling software developed for this purpose. The significance of commissioning and inhabitant engagement is highlighted, and the crucial question of 'how much is enough?' is tackled in relation to modelling and feedback. Chapter 6 examines the notion of housing life cycles and the role of longitudinal BPE studies to improve design, organisational learning and reflective practice.

3. *Training for feedback*

   This section describes the essential BPE methods and processes needed to carry out successful housing BPE studies routinely. Chapter 7 outlines international sustainable design standards for benchmarking, and what BPE methods to use when, using a 'drill down' approach. Chapter 8 illustrates new directions in housing occupancy feedback, covering prototyping, anthropology, ethnography, digital media, action research and institutional innovation. Chapter 9 describes BPE educational methods and training devised and delivered by the author and others for students, professionals, clients and inhabitants in the UK. The chapter concludes with a description of the personal attributes and skills required to become a successful housing occupancy feedback practitioner.

4. *Application and case studies*

   The key housing BPE challenges are discussed in this section. Three international newbuild and retrofit BPE case studies in Chapter 10 describe particular contributions to the development of housing BPE methodology, as well as their impact on policy development. Chapter 11 briefly sets out the historical construction, contractual and legal context for housing design and retrofit in the UK, before discussing two relevant BPE case studies. Chapter 12 details the costs associated with different scales of BPE in practice, providing a critical cost-benefit analysis of three key studies undertaken by the author to help housing stakeholders identify best value for particular projects.

5. *Challenges for the future*

The penultimate section of the book addresses two critical challenges for further BPE development: ethics and feedback. Chapter 13 sets out a vital ethics process, highlighting key issues such as privacy, ownership, metadata analysis and data protection. This leads into the most effective and practical way to build BPE feedback into organisational systems linked to the RIBA Plan of Work process, as covered in Chapter 14. This ensures that any vital lessons are effectively communicated to all relevant housing stakeholders. Chapter 15 concludes by summarising the main barriers and opportunities for housing performance evaluation. It revisits some of the key lessons highlighted in the book, and presents a forward-looking agenda with 10 key questions for policymakers to consider in terms of making housing BPE an integral part of the design and build process.

6. *Housing feedback primer*

The aim of the compact primer is to give you a simple practical approach to conducting a housing BPE study, and explains what should happen afterwards to make a difference. It describes what do in relation to each of the key principles and methods described in the book, as well as when to use them. References back to the relevant chapters in the book provide more information on the underlying reasoning and detail behind the various tasks described in the primer.

Route map for different users

The table on page xiv suggests which chapters are most relevant for each group of stakeholders in the built environment. The route map is based on the assessment of working with various stakeholders over the years, and noticing what interests them most. For practitioners in a hurry to start doing BPE studies, go straight to chapters 7, 12, 13 and 14, or just jump to the primer. You may disagree with the signposting, and prefer to find your own way. You are heartily encouraged to do so, as there are many ways to read this book.

ROUTE MAP FOR READERS OF THIS BOOK	Stakeholders					
Book content	Architect/ Engineer	Surveyor/ Manager	Client/ Developer	Policymaker	Academic/ Researcher	Inhabitant
Introduction	H	H	H	H	H	H
BACKGROUND						
1: History	M	L	L	H	H	L
2: Drivers	M	M	H	H	M	H
LEARNING						
3: Physical theory	M	M	M	M	H	L
4: Socio-cultural theory	M	M	M	M	H	L
5: Modelling and reality	H	H	M	M	M	L
6: Longitudinal feedback	M	H	H	H	M	L
TRAINING						
7: BPE techniques evaluated	H	H	M	M	H	H
8: Innovation in feedback	H	H	M	M	H	H
9: Education for feedback	M	M	M	H	H	M
APPLICATION/CASE STUDIES						
10: The international context	H	H	H	M	M	M
11: The UK context	H	H	H	M	M	M
12: Costing housing BPE	H	H	H	M	M	H
CHALLENGES						
13: The ethics of feedback	M	M	H	H	M	H
14: Effective feedback loops	H	H	H	H	H	M
15: Next steps	M	M	M	H	M	L
Housing feedback primer	H	H	H	H	H	H

H = high interest   M = medium interest   L = low interest

SECTION

1

BACKGROUND

ONE
A SHORT HISTORY OF HOUSING EVALUATION

" You cannot deal with the people and their houses separately. **"**

Octavia Hill in
Macmillan's Magazine,
XXIV October 1871, p 464

This chapter outlines a brief history which charts the international development of housing performance evaluation over the last century – from public health work in the 19th century, via building science departments in the 1960s, through its exploration in the social sciences, right up to the latest interdisciplinary ventures.

The following aspects are included in this chapter:

- Victorian housing: a public health warning
- Building science and housing monitoring
- The occupant survey
- Environment behaviour and post-occupancy evaluation
- Participatory design
- Building performance evaluation
- The relevance of feedback for housing performance evaluation
- Towards a hybridity of evaluation: interdisciplinary collaboration

Victorian housing: a public health warning

Early housing performance legislation in the UK

The ancient Vitruvian building principles[1] of *firmitas* (being durable), *utilitas* (functioning well) and *venustas* (being delightful) remain key design goals for housing. Despite this, numerous housing disasters have occurred over the centuries. The Great Fire of London in 1666 destroyed the homes of 70,000 inhabitants[2] and resulted in the London Rebuilding Act of 1667. This was the first attempt in the UK to legally limit the spread of fire and prevent the building of dangerous structures, later consolidated in the Building Act of 1774, which influenced housing standards for the next two centuries. In the 19th century, local medical authorities realised that there was a connection between the insanitary physical conditions of homes and the poor health of inhabitants, through the evidenced-based approach of the leading British physician John Snow (1813–58) and others. These conditions related to dampness, overcrowding, poor heating, sanitation, light, water and air quality. Adequate sewage drainage was required to prevent disease, and this was enshrined in the Metropolis Management Act of 1855. The Sanitary Act (1866) then made overcrowding in housing an offence, while the Artisans' and Labourers' Dwellings Improvement Act of 1868 enabled the demolition of insanitary housing to prevent the spread of disease from poorer neighbourhoods through wider society.[3] Finally, the Public Health Act of 1875 required local authorities to provide sanitary housing through a series of by-laws. Ironically, this early performance-based legislation was the precursor to a long line of regulation to improve housing design, but not necessarily its performance in reality, as there was no provision for performance evaluation.

Housing feedback and inhabitants' behaviour

With Victorian housing for poorer inhabitants in the UK increasingly regarded as a public health hazard (see Figure 1.1), philanthropic model dwelling companies developed in the second half of the 19th century to re-house inner-city slum inhabitants. The secretary of one company based in London, the Metropolitan Association, even went so far as to advertise the evidential healthiness of their housing estates in terms of sanitary provision in an attempt to break down prejudice against the novel housing design of high-density housing blocks[4]. This was an early form of ad-hoc housing feedback procured to help the owner promote their product. Victorian housing reformers such as Octavia Hill also made a connection between the quality of houses and the character and habits of the people living in them. They believed that housing would

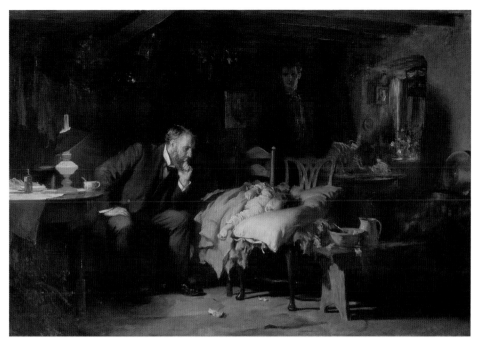

Figure 1.1 Victorian housing reformers linked inhabitants' health to good sanitation

soon fall into disrepair and ruin if the inhabitants were not 'reformed'.[5] Interestingly, some contemporary housing performance evaluation now also aims to understand inhabitants' motivations, expectations and behaviours in relation to housing occupancy, making a similar connection between housing and behaviour, but sometimes falling into the same rhetoric of inhabitant 'reform'/blaming as discussed in Chapter 4.

Building science and housing monitoring

Modernism meets measurement

After the devastation of the First World War, there was a renewed interest in healthy housing led by the modern movement in architecture. This was evidenced through Richard Neutra's Health House for Dr Lovell in the USA, Le Corbusier's manifesto *Towards a New Architecture* in France, and demonstrated in the Weissenhofsiedlung housing settlement in Stuttgart, Germany, in 1927.[6] The world's first organisation to undertake evidence-based building research, the UK Building Research Station (BRS), was set up in 1921 in West London, and by 1926 had grown to over a 100 staff.[7] Experimenting and testing of materials at this time provided some random empirical evidence of housing performance in terms of its fabric, but little systemic study to show how the housing was performing overall and what the inhabitants thought of it. One exception was 358 measurements of the air change rate in six homes in London using the decay of coal-gas released into the home, in 1943.[8] This study concluded that liberal window openings could provide enough ventilation in these homes, given all the flues, gratings and cracks present.

Following the huge destruction of the Second World War there was a need to produce mass housing quickly in Europe. The UK Housing Act in 1949 enabled local authorities to build homes for all classes. Local authority architects were also highly influenced by Le Corbusier's call for a new 'science of housing'[9] and an extensive drive for prefabrication, with government subsidies given to encourage new types of non-traditional housing. This was exemplified in Sheffield's famous Park Hill development, where 'Dimensions, light and layout were all measured so as to ensure that each resident received a regulation flat to provide for their basic needs.'[10] However, UK housing was still not fit for purpose, with the government's Fuel and Power Advisory Committee reporting in 1946 that heat insulation in the construction of homes was neglected given that '… in our inconsistent climate, space heating is required at most times of the year'.[11] Postwar studies carried out by the BRS developed the first mathematical models for measuring lighting, acoustics, and thermal and ventilation performance in housing, backed up with testing in experimental houses.[12] This informed design guidance in UK government housing manuals in the 1940s, and eventually informed housing standards in the UK from the 1960s onwards.

Monitoring in housing comes of age

The first 'English House Condition Survey' took place in 1967, examining fitness for purpose, disrepair, and the availability of basic amenities such as sanitation and heating. This was perhaps the first routine evaluation of the existing building stock in the UK. The RIBA also called on its architect members to undertake performance feedback on completed building projects in Part M of its first detailed Plan of Work, published in 1964. The Building Performance Research Unit led by Tom Markus in Strathclyde University, Glasgow, then produced the first definitive text on *Building Performance* in 1972, which presciently included a model of how to potentially undertake systematic housing performance evaluation showing technical and cost components alongside environmental and behavioural perspectives[13] (see Figure 1.2). This was quite radical at the time.

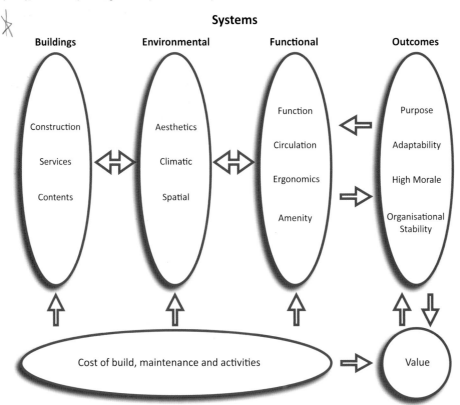

Figure 1.2: Early model of housing performance evaluation

Following the global oil crisis of 1973–4, the UK government started to systematically document how much energy was being used in homes, with the renamed Building Research Establishment (BRE, formerly BRS) expanding to 1350 staff by 1975 to carry out work on housing energy conservation measures.[14] Throughout the 1980s and 1990s the BRE was at the forefront of many UK government-funded 'best practice' housing demonstration projects specifically to develop energy conservation and energy efficiency, alongside numerous European-wide housing demonstrator projects such as CADDET. While these initiatives monitored energy in use, they generally lacked input from the inhabitants, and any broader evaluation of ventilation, water use or air quality. They were not housing POE studies in the full meaning of the term, and others had to develop a more holistic approach.

The occupant survey

Mass Observation begins

With nearly four million homes damaged after the Second World War, the UK government undertook a bold experiment to understand what kind of housing its citizens would like, drawing on national Mass Observation studies which had examined people's lives in their homes in great detail during the war.[15] The director of these Mass Observation studies noted:

'It is the experts who choose and build the house; it is the ordinary husband and wife who choose and build the home. Often the housemakers and home-makers plan their houses and homes separately, each with little idea of what the other is getting at and looking for. Mass Observation's job is to provide the link between expert and amateur, planner and planned-for, the democratic leader and the democrat.'[16]

This careful differentiation between 'house' and 'home' is important, as these terms (along with 'housing', 'household', 'dwelling') are used interchangeably in many housing BPE studies without noting the difference. The former term tends to privilege physical and quantitative measurements, while the latter includes interpretations of social and cultural expectations and norms, as well as emotions and relationships.[17] In this book, I favour the word 'home' over 'house' or 'dwelling', in order to capture this broader dimension of housing BPE. At the same time, it is important to situate a 'home' within a societal 'housing' context.

The wartime Mass Observation report covered 1,100 detailed interviews[18] and some results did eventually make their way into the 1944 the government report 'Design of Dwellings' (known as the Dudley report).[19] This was perhaps the first example of housing design guidelines informed directly by inhabitants' feedback.

The BRS, which was mainly wedded to materials studies, branched out into human factor studies during this period.[20] In 1942, a study of a small development of modern flats (Kensall House, designed by Maxwell Fry) revealed that 61 of the 68 residents surveyed were well satisfied with this type of accommodation.[21] The London County Council also carried out housing surveys of pre-1914, interwar and postwar council housing.[22] In 1967 the Ministry of Housing and Local Government also began its survey on 'The Estate Outside the Dwelling', which included investigating the effects of high buildings and subsequent tenant satisfaction.[23]

The limitations of pre-structured surveys

The objective of the above surveys was to discover what worked well and what did not. Unfortunately, the inhabitants had to provide answers within a framework that suited the architect, estate manager and town planner. The focus on measurable criteria only, such as the size of various rooms, the amount of play space for children or the amount of sunlight rooms received was clearly intended to provide statistical feedback that could be used by planners in the future. Yet the surveyors were unable to produce coherent results, with '… no clear evidence to suggest that some building forms were more satisfactory for some household types than others'.[24] From the late 1970s onwards, planning and design evaluation was generally carried out without professional survey teams. Despite this, housing organisations continue to perform tenant satisfaction surveys, largely for management rather than design purposes. A significant gap remained between the physical monitoring of housing and the separate activity of tenant surveys in the UK, with neither providing a satisfactory account of *why* homes were performing in the way they did.

Environment behaviour and post-occupancy evaluation

Social sciences enter the field

One innovative attempt to go further than separate passive tenant surveys and physical monitoring exercises occurred at the new Park Hill Estate in Sheffield, where the city council employed a sociologist, Mrs J.F. Demers, to become the first resident in 1960 and provide inhabitants with expert guidance, encouragement and education. The Sheffield Housing Committee bravely stated: 'It must be left to others – particularly the occupants – to judge to what extent the Architects have been successful in solving this social problem'.[25] This was a reality check, recognising the need to go more deeply into the contextual issues surrounding housing performance in order to understand it.

A new field of environmental behaviour promised another way forward, related to people's perception of space, as set out in Edward T. Hall's book *The Hidden Dimension*.[26] This followed on from seminal studies in the USA by Robert Sommer in 1959, specifically on personal space that provided a more theoretical basis for understanding housing performance as a relationship between people and the built environment. The approach was taken up in the UK by the psychologist David Canter in his books *Psychology for Architects*[27] as well as *Psychology and the Built Environment*[28] (co-edited with Terence Lee). Environmental behaviour studies continued to develop in the USA where architects such as Henry Sanoff[29] and sociologists such as John Zeisel were interested in inhabitants' responses to public housing.[30] It was these researchers and practitioners who led the way in promoting the theoretical idea of 'post-occupancy evaluation' (POE),[31] which combined numerous methods to overcome the gap between physical housing monitoring studies and socio-cultural evaluations of the home.[32] John Zeisel's book *Inquiry by Design*[33] became a key text in this area. Clare Cooper Marcus and Wendy Sarkissian[34] also carefully evaluated over 100 post-occupancy housing studies, using this approach.

The first formal definition of POE methods as an approach

The first international book directly concerned with formalising POE methods covered three key stages: indicative, investigative and diagnostic (with increasingly complex methods at each stage) and three categories of performance criteria: people, settings and relational concepts.[35] By this time, POE studies had already burgeoned around the world. Most of these studies, however, concerned public and commercial buildings such as schools, hospitals and offices, rather than housing, and virtually no POE was carried out on individual houses at this stage.[36]

Participatory design

Who is really leading whom?

Despite studies capturing how inhabitants responded to their homes, developers and built environment professionals still felt that inhabitants did not grasp the benefits of future housing design propositions. As a result, urban theorists introduced techniques that architects and planners could use to the present their plans to people in order to help them think like a designer or planner, and engage inhabitants upstream of their later response to the completed development.

As early as the 1950s, practices such as holding public meetings, interacting with communities and being visible on proposed sites of redevelopment were recommended as techniques to encourage people to evaluate their needs more 'rationally', and thus hopefully support the proposed scheme: '... if people can be persuaded that a need is real and important – that is if they understand the why of the matter – then what to do about it, or the how of activity called for can generally be managed'.[37]

The cognitive framework for inhabitants to understand the proposed design of their home is once again here professionally predefined. The designer sets the agenda and terms of reference and then asks the inhabitants to respond to them, rather than enabling inhabitants to question them in the first place, and devise their own terms of evaluation. The assumption is that the designer's viewpoint is more 'rational' than that of the future inhabitants, who need to be 'educated' into adopting the designer's view.

As early as 1975, researchers[38] were already asking for genuine unmediated participation by inhabitants: 'Unless participation is acted out thoroughly at an early stage, those who emphasise the need for sociological research and high-quality design are still missing *the* basic point and advocating housing designed *for* people instead of *by* people.'

Learning how to listen

Participatory planning and design has strong roots in the USA community protest movements centred around maintaining affordable housing and its link to 'programming'. This is essentially a systematic activity undertaken by designers to gather detailed information about their developer-clients' and occupants' needs, and thus set out an appropriate design direction. Interestingly, Henry Sanoff made an early professional connection between feedback from participatory planning and POE, pointing out that 'Learning how to listen, not only to the paying client, but to the people who use and are effected [sic] by the environment, within the social and historic context, can produce a professional with an expanded capacity for shaping the future.'[39]

Community Design Centres sprang up in the USA as a result of this new attitude, advocating for community housing needs at a local level. One outcome from this was Sherry Arnstein's famous Ladder of Participation, which outlined the different levels of community empowerment that existed according to the method of participation used.[40] This 'ladder' illustrates the degree to which housing POE studies empower inhabitants, either as 'subjects' with little control over the POE results, through to co-producing 'participants' with full control over their housing situation (see Figure 1.3). Sanoff also realised the need to plan for evaluation of inhabitants' needs, activity and satisfaction, at any stage of the building life cycle, right at the beginning of a project as part of the participatory design process. However, until the 1970s energy crisis hit, most USA housing evaluation studies were only concerned with social, political and cultural aspects, with little or no attention paid to the actual environmental impact of the development itself.[41]

Community architecture and inhabitant engagement

Meanwhile in the UK, the 1970s and 1980s saw a parallel development of interest in participatory planning and design through an emerging 'community architecture' movement.[42][43] This was in response to a perceived housing crisis and the failure of many of the more recent UK council housing initiatives. The movement was supported in 1987 by the Royal Institute of British Architects under its then-president, Rod Hackney, and promoted by Prince Charles.[44] Activists promoted self-build and self-help housing, where the occupants literally help themselves to build or renovate their own homes, and were strongly

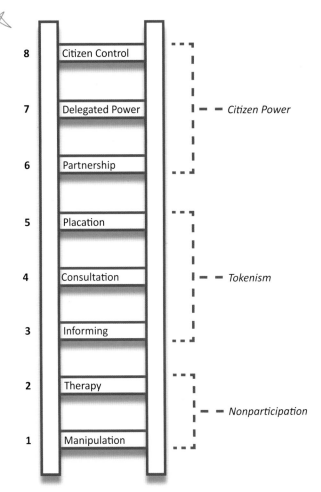

8 Citizen Control

7 Delegated Power — Citizen Power

6 Partnership

5 Placation

4 Consultation — Tokenism

3 Informing

2 Therapy

— Nonparticipation

1 Manipulation

1 **Manipulation** and 2 **Therapy.** Nonparticipative, cure or educate the participants. achieve public support by PR.

3 **Informing**. One way flow of information.

4 **Consultation**. Attitude surveys, neighbourhood meetings.

5 **Placation**. Allows citizens to advise but retains for power holders the right to judge the legitimacy or feasibility of the advice.

6 **Partnership**. Power is redistributed through negotiation between citizens and power holders. Shared decision-making responsibilities.

7 **Delegated Power** to make decisions.

8 **Citizen Control**. Participants handle the entire job of planning, policy making and managing a programme.

Figure 1.3 Sherry Arnstein's Ladder of Participation

influenced by John Turner's radical housing work in Peru.[45] Architects such as Colin Ward[46], Ralph Erskine and Walter Segal[47] all promoted different ways for inhabitants to engage with innovative housing management, design and build processes, but with different levels of empowerment. Ward advocated that people should take control of the housing process for themselves through occupying undeveloped land and building homes for themselves, without employing professionals. Erskine offered future inhabitants a chance to discuss their concerns and ideas with him and his design team on site, related to his proposed Byker Wall housing redevelopment in Newcastle. Segal gave people the opportunity to build their own homes, but based on his own strict design proposals known as the Segal Method. At the same time a number of Community Technical Aid Centres sprang up across the UK, and these had affiliated into an association with over 50 member organisations by 1985.[48]

The evolution of community architecture, and subsequently more entrepreneurial community development in both the USA and the UK, resulted in only a small number of new or retrofitted housing developments which were genuinely influenced by the inhabitants themselves at the design stage. This was because, despite regulatory requirements in the USA and substantial initial funding from governments in both countries, the institutionalisation of participatory design undermined its own emancipatory aims. The creation of narrowly defined participatory design guidance and restrictive terms and conditions that communities trying to obtain funding had to adhere to effectively made participatory design a 'top-down' process once again. The judging of the developments was in professional aesthetic terms that disparaged non-conformists, leading to disillusionment on all sides. Once UK government funding ceased in the 1980s, the movement retreated into a small, persistent margin of activity, often linked with housing cooperatives or housing associations, which were forced to find funding where they could. Theoretically, participatory planning and design in the UK could have led to the same sophisticated evaluative studies that were occurring in the USA to help understand inhabitants' needs, but in reality these did not happen beyond the usual cursory 'customer satisfaction survey'.

Building performance evaluation

Linking outcomes to intentions
In the early 1990s, my own professional experience as a practising community housing architect working in the Glasgow community-based architecture cooperative ASSIST,[49] led to the realisation that my first housing design solutions for retrofitting 1960s housing developments using participatory design methods were not working. The inhabitants were not 'behaving' as they were supposed to in the newly built south-facing balcony conservatories, and were trying to use these as all-season living rooms in order to gain extra space (see Figure 1.4).

Two personal insights developed over the next few years: first, that I needed to learn more and write guidance on sustainable housing design, working with an experienced housing client and planner; and, more specifically, that I needed to learn how and why inhabitants used their new homes in the ways in which they did. The resulting book commission from the Scottish government agency at the time proved both an epiphany and a journey. Thirteen housing case studies examined and compared how the homes were performing after completion in relation to the original design aims, covering a wide range of factors and delivering a broad set of recommendations for the design team as result.[50] The book was later re-commissioned by the same agency, and substantially revised as a free online resource with a total of 23 housing case studies, published in 2007.[51]

Figure 1.4 Balcony conservatories used as extended living rooms lose heat

Wolfgang Preiser now moved on from his pioneering definition of POE methodology. He developed the term 'building performance evaluation' (BPE) as a broader, systematic process which placed POE within an evaluation of the complete life cycle of a building from inception through to occupancy, to ensure that this knowledge was fed back into the design cycle[52][53] (see Figure 1.5). This new concept strongly influenced the development of several major UK government-funded innovative housing design research programmes that ran from 2009 to 2015 and included BPE studies – 'Retrofit for the Future',[54] 'AIMC4'[55] and 'Building Performance Evaluation'.

Resistance to housing BPE in the UK

Developing housing performance evaluation more widely in the UK was a struggle, as clients remained reluctant to share their results collectively. In 2002, a regional office of the Scottish government housing agency commissioned me to undertake a full POE of an innovative all-timber housing development

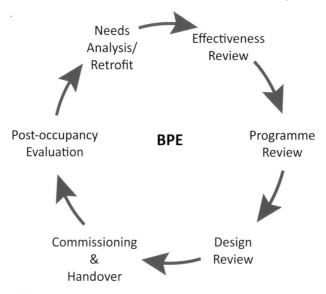

Figure 1.5 The BPE life cycle

in Kincardine O'Neil in Deeside, Scotland (see Figure 1.6). For the first time, I experience first-hand the institutional resistance to POE that exists when clients feel they will be blamed for mistakes. The report and its recommendations for the future design of affordable housing in Scotland were not published. Fortunately, several years later, a researcher in the same agency stumbled on the report on a shelf, and immediately commissioned a further report to update the findings with an extended POE study.[56] After presenting the findings of this report to an international conference in 2004, I was invited to set up a housing branch of the Usable Buildings Trust, a leading UK organisation promoting POE. In 2007, the UK Good Homes Alliance (GHA) became the first house builders' membership organisation to insist that developers openly monitor their homes for performance during occupancy. At the time of writing, however, the GHA have dropped this mandate and housing BPE processes have yet to be legislated in the UK. The POE coverage of the UK housing stock remains patchy at best, despite the need for good feedback on housing performance being more important than ever.

Figure 1.6 Kincardine O'Neil – an all-timber housing development, including shingle roof

The relevance of feedback for housing performance evaluation

A classic dictionary definition of 'feedback' is as follows: 'Information about reactions to a product, a person's performance of a task, etc. which is used as a basis for improvement (Oxford Living Dictionary). BPE aims to make feedback routine in housing projects as '… a way of quality control in the more repetitive projects; as a necessary part of hypothesis testing in innovative ones; as a means of increasing awareness of chronic problems, changing requirements and emerging properties; and as a way of promoting fine-tuning and team learning'.[57]

The feedback that housing developers generally receive on the performance of their new projects are a narrow range of customer satisfaction comments from inhabitants and the need to attend to sporadically reported problems within the warranty of the contract. There is usually no performance feedback and learning as a formal organisational process. Too often, developers solve problems on an individual basis without looking at all the underlying causes. Emerging technical and social problems in housing development have become chronic through lack of good, comprehensive, feedback. Typical examples of this are the poor levels of airtightness achieved, until testing became mandatory, and energy bills that are often three times more than the design target.

The development of BPE processes in the last decade means that housing organisations can now undertake thorough investigations of their product development as well as using BPE to fine-tune the delivery and performance of the product as built and inhabited. These processes are discussed in Chapters 6 and 7.

Towards a hybridity of evaluation: interdisciplinary collaboration

The promise of interdisciplinary housing BPE

The final ingredient that concludes this brief review of housing BPE development is interdisciplinary collaboration. Early housing energy monitoring was undertaken by engineering researchers who were interested in building physics, with little interest in finding out how the inhabitants were affecting the results. Designers and social scientists rarely interacted with these engineers, and ended up developing their own social evaluation studies, but without any physical monitoring. An effective BPE study needs both sets of studies triangulated to provide a socio-technical analysis of housing performance. The case study method provides an excellent means of integrating these separate approaches.[58] Undertaking an interdisciplinary BPE study can be expensive in labour costs, however. One solution is to use BPE specialists who have some expertise in both social and technical approaches and are able to carry out the initial diagnosis[59] with the use of additional disciplines if required beyond this stage.

The obstacles to interdisciplinary BPE

There has been a degree of antagonism between social scientists, designers and engineers when it comes to developing BPE methodology, with some claiming that it is impossible to combine these three approaches as they are incommensurate, and that instead they should sit side by side as part of any socio-technical study.[60] Architects can span engineering and social science disciplines as part of their holistic appraisal of design needs and processes. As practice-based researchers, they can act as a bridge between the two sets of approaches, providing they can overcome their own professional self-interest. To do this requires a 'new professionalism' – one that is more ethically minded and fully engaged with sharing building performance knowledge more broadly.[61] Genuinely interdisciplinary collaboration and understanding is vital for housing BPE to be able to include people as well as buildings.

TWO
DRIVERS FOR BUILDING PERFORMANCE AND OCCUPANT FEEDBACK

" It's a line in the sand and what it says to our species is that this is the moment and we must act now. **"**

Debra Roberts, co-chair of the IPCC working group on impacts, 2018

What, then, are the key challenges facing housing design and development today in relation to feedback on performance? This chapter provides the socio-political context for this book, outlining key drivers for housing performance evaluation, including climate change, health, resource use and planning legislation, the Energy Performance Directive and other European Union (EU) initiatives. The chapter covers the following aspects:

- Carbon positive housing
- Resource use
- Adapting to climate change
- Energy, carbon emissions, and building policy and regulations
- Healthy homes and neighbourhoods

Carbon positive housing

Climate change impacts on and from housing

Our climate is changing for the worse. Scientists around the world attribute much of this change to the rapid increase in human-made carbon emissions that has occurred over the last century, and which continues on an upward trajectory.[1] Without radical measures to rapidly reduce these emissions, we are likely to see drastic increases in catastrophic temperature rises, heatwaves, flooding and storm damage. We are already seeing these changes today (see Figure 2.1). What role does housing BPE have in dealing with this?

In terms of building development, housing is the single largest emitter of all CO_2 emissions in the world.[2]

Figure 2.1 Three concurrent hurricanes in the Gulf of Mexico

Households also represent around 19% of all greenhouse gas end-user emissions in the EU, with over 800 million tonnes greenhouse gas equivalent CO_2 emissions per year.[3] Worryingly, BPE shows that housing routinely emits more than 2.5 times the amount of carbon compared with design predictions in the UK.[4] Perhaps not surprisingly, the Intergovernmental Panel on Climate Change (IPCC) sees the need for improved building performance as a critical issue in terms of reducing carbon emissions and climate change.[5] A number of responses to this from the housing sector drive performance evaluation.

Nearly zero, zero and carbon positive housing

In 2007, the UK government introduced its 'Code for Sustainable Homes' as a voluntary standard to continuously improve home building. The government's ultimate aim was for all newbuild homes to generate enough renewable energy on site to offset the amount used for space heating, hot water, lighting and ventilation by 2016. Many saw this approach as demanding 'too much, too quickly' given the relatively undeveloped market. Nevertheless, house builders and designers positively engaged with the agenda and developed costed solutions ready to roll out in 2016. In 2015, however, the UK government unfortunately scrapped this pathway towards 'zero carbon' housing.

In a more pragmatic move in 2010, the European Commission had mandated that all newbuild homes within the EU should be 'nearly zero energy buildings' (nZEBs) by 2020, under its Energy Performance of Buildings Directive (EPBD). This was updated in 2018 to accelerate the cost-effective renovation of existing buildings, aiming for a fully decarbonised building stock by 2050, with appropriate indoor air quality and comfort. The definition of 'nearly zero' in the EPBD is difficult to measure for compliance. Many designers and authorities have thus adopted the more prescriptive Passivhaus standard as a more concrete measure of compliance, with its requirement for $120kWh/m^2$ year maximum used for heating and cooling the home.

As a driver for housing BPE, the Passivhaus standard provides a number of important parameters, including thermal bridging, overheating, solar gain, renewable energy, and retrofit energy standards. Surprisingly, there is no Passivhaus requirement for BPE to date, although basic monitoring frequently takes place in these projects.

These housing standards are not addressing climate change quickly enough, however, and various countries are now pioneering 'carbon positive' or 'net zero+' housing solutions. These aim to generate enough renewable energy from the individual home to offset any carbon emissions from that home and give extra energy back to the grid to help offset carbon emissions elsewhere. There are no agreed standards for this type of project yet. All of these solutions are in danger of ignoring the resources needed to make homes in the first place, however, which can be significant.

Resource use

Why energy is not a resource

One definition of 'resource' is 'a source of supply, support, or aid, especially one that can be readily drawn upon when needed' (http://www.dictionary.com/browse/resource) and this relates to assets that can be drawn on to provide housing. Resource use is another critical driver for housing BPE.

Energy, however, is not a 'resource' in itself, but an enabling *medium* harvested from natural resources. There are two types of energy sources for homes – renewable and non-renewable – and the challenge for BPE is to evaluate whether these are beneficial or not in terms of their overall impact. Housing BPE should assess the energy and carbon emissions embedded in the harvesting process, and the efficiency of the use of the energy sources in the home. However, the overriding issue is evaluating how much we actually need as humans within the constraints of our planetary resources – what is *sufficient* energy for our homes[6] and what constitutes a *reasonable* level of services. This overall approach should be a major driver for housing BPE, but we are not there yet.

Water evaluation

Water metering is required for new homes in the UK. This is particularly important in relation to energy use, given that one cubic metre of urban drinking water typically requires around 1 kWh to produce in the UK. Year-round usable water is also a limited resource in many countries, given the increasing flood/drought cycles that come with climate change. Household water use varies hugely, depending on the number of people in a home and their personal needs. In the UK, a single-person household used around 50 cubic metres annually in 2018, while a five-person household used 191 cubic metres.[7] Around a third of UK households use drinking-quality water to flush toilets, a quarter for personal washing, and a fifth for washing clothes.[8] These three activities use up over 75% of household water. Water use is as important as energy use. It is now routinely assessed as part of BPE methodology and is a key driver[9] (see Figure 2.2).

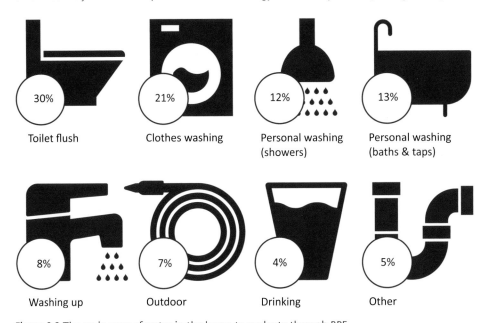

30%	21%	12%	13%
Toilet flush	Clothes washing	Personal washing (showers)	Personal washing (baths & taps)
8%	7%	4%	5%
Washing up	Outdoor	Drinking	Other

Figure 2.2 The main uses of water in the home to evaluate through BPE

Materials evaluation

Materials are essential for making homes. Generally, relatively unprocessed minerals and vegetation such as stone, clay, wood, bamboo and straw require less energy per weight to convert into basic construction products than artificially and highly processed and mixed compounds such as concrete, steel and plastics. This embodied energy, and the associated embodied carbon emissions, varies depending on the manufacturing, transportation and construction processes used. Scarcity, pollution, toxicity, recycling, waste and durability are all aspects to take into account when evaluating construction materials as part of their life cycle assessment (LCA). Embodied energy and carbon emission rates are often a reasonable proxy for these factors when LCA is not possible due to its complexity and cost.

As homes become more energy efficient, the amount of energy and associated carbon emissions embodied in them becomes increasingly important and can even exceed the amount of energy in use/ carbon emissions over a number of years.[10] Despite calls for the regulation of embodied carbon emissions[11] there is still not a robust open-source method and database to use for these purposes, with few BPE studies reporting embodied energy factors compared to energy in use.[12] The evaluation of material use is still not a strong driver for BPE studies, though it clearly should be.

Recycling and reuse evaluation

Recycling or reusing resources in the construction, operation and maintenance of the home can save energy as well as minimising the extraction of limited resources. While there is no regulatory requirement to use either reused or recycled construction materials or water in housing, there are now stringent regulations on the installation of water recycling systems related to housing in many countries, which should act as a driver for housing BPE. However very few BPE studies actually examine these types of water systems or how households reuse their water.[13]

Regenerative housing and renewable resources

Positive development aims to give back more to a place and its communities than was there originally in terms of social, environmental and economic resources. This concept has been extended into 'regenerative development', where buildings are co-created with communities in relation to their local ecosystems.[14] This is one step beyond expanding building performance parameters to include biodiversity and ecosystem services and is not yet a major driver for housing BPE, though again, it should be.[15]

Adapting to climate change

Mitigation

Climate change mitigation means action to minimise the rate and magnitude of long-term climate change by reducing anthropogenic greenhouse gas emissions, or increasing the means to absorb carbon dioxide (e.g. through reforestation). The UK introduced a mandatory Climate Change Act in 2008, committing successive governments to achieving an 80% overall reduction in these emissions in the UK by 2050 (using a 1990 baseline) which was amended to 100% in 2019. In 2012, the UK's Committee on Climate Change demanded a 50% reduction in carbon emissions from the housing sector by 2030, which equates to a 38% reduction in space heating and hot water demand.[16] The EU has since suggested that buildings will need to reduce their carbon emissions by 90% by 2050.[17] Consequently, one report strongly advocates the radical move of banning of all fossil fuel heating systems by 2026 to help meet this goal.[18]

Key drivers for climate change mitigation in the building sector include reducing energy use, increasing energy efficiency and the use of renewable energy to 'decarbonise' the national power grid and the home.[19] The measurement of carbon emissions related to energy use is critical for evaluating climate change mitigation in housing BPE studies.

Adaptation

Climate change adaptation is 'the adjustment in natural or human systems in response to actual or expected climatic stimuli or their effects, which moderates harm or exploits beneficial opportunities'.[20] Homes need to be adapted to cope with the negative impacts of climate change as well as enhancing their ability to take advantage of any benefits. Housing details now need to cope with exceptional storm conditions, including excessive cold, snow, wind and monsoon-level rainfall events as well as severe flooding, heatwaves, drought and generally rising temperatures (see Figure 2.3). Moving from reactive to advanced planning for this type of adaptation is particularly challenging for households without further government support.[21]

Figure 2.3 Homes need adaptation capacities to cope with increasing storm levels

Recent BPE studies show that significant overheating of new homes is now occurring in the UK during the summer, particularly in urban areas. This is due to a combination of increasing external air temperatures as well as the increased airtightness of these homes.[22] While housing BPE studies are now driving some of the developing adaptation strategies in relation to overheating, many climate change adaptation drivers are not yet translated into mainstream housing BPE methodology, particularly in relation to flooding, drought and storms, or the degree to which homes are futureproofed against climate change.

Resilience

Housing performance depends on robust construction and services as well as built-in resilience to cope with future changes to our climate.[23] At present, there are no building regulatory requirements for new or retrofitted homes to be resilient in relation to the worsening climate, or to anticipate future conditions to avoid obsolescence. New homes are designed to last for at least 60 years in the UK, so this is a significant omission. Planning requirements stress the need to design for predicted flooding, but not to cope with the accompanying droughts that now inevitably occur in the same flooding cycle in the UK. Both robustness and drought considerations are yet to translate into BPE methodology.

As uncertainty increases in relation to the rapid increase in climate disruption, another key resilience driver is 'redundancy' – the means to provide a single housing function through multiple means. Typically, this is linked to different ventilation and heating options, so that if one means fails, there is always another one available.[24] The move towards 'smart' housing is in danger of eliminating traditional redundancy, such as openable windows and decentralised heat sources, in favour of centralised mechanical systems, where if one part fails, it all fails. Having alternatives available provides additional agency and knowledge for inhabitants that is empowering and helps to build their capacities. Redundancy should therefore be another key driver for BPE housing methodology.

Futureproofing
Weather datasets are now available which predict the future climate, both globally and locally, and should be part of regulatory compliance in relation to overheating. This would help us to construct and retrofit homes that are effectively futureproofed through this compliance and better adapted to cope with the changes. Futureproofing should be another a key driver for BPE methodology, although it is not at present.

Energy, carbon emissions, and building policy and regulations

Metric variations
There is wide variation of energy efficiency and carbon emission levels in building regulations globally, which takes account of different climates and cultures. This, however, makes it difficult to internationally compare the energy and carbon performance of homes, which needs to be taken into account in BPE methodology.[25] Different countries are at different stages of building regulation development for housing, although most have now moved to performance-based codes rather than prescriptive ones.[26]

Compliance: measurable, reportable, verifiable
The EU EPBD regulations set clear energy performance targets. But without actual verification procedures in place, and given the routine gap between actual and designed energy use, there is no guarantee that these targets are meaningful.[27]

Under these regulations, housing developers are required to provide an Energy Performance Certificate (EPC) to all prospective buyers and tenants when a building is for sale or rent. The certificate must include an energy performance rating as well as recommendations for improvements. In the UK the credibility of these certificates has been significantly undermined due to the high level of variation of results from different assessors[28] and due to varying assessments of the heating efficiency, thermal performance of the exterior walls, and geometry of homes.[29] As such, EPCs in the UK do not yet provide a credible driver for BPE methodology in terms of measurable verification.

Modelling and reality
All homes are subject to Part L1A (Conservation of fuel and power in new dwellings), and Part L1B (for changes to existing dwellings) in the English building regulations. These cover the energy efficiency of the exterior fabric and the services providing heating, lighting, hot water and ventilation as regulated energy loads. Any proposed work has to meet a set Target Fabric Energy Efficiency (TFEE) rate and Target CO_2 Emissions Rate (TER) and the Standard

Assessment Procedure (SAP) provides the underlying model and calculations for working out whether or not a home complies in theory with these regulations. However, for BPE purposes, SAP is unreliable as it makes many assumptions, is overly simplistic and yet is still easy enough for assessors to get wrong, as recent housing BPE studies have consistently demonstrated.[30]

The regulations also include the need for appropriate commissioning of key services such as space and hot water heating as well as ventilation, and the provision of adequate guidance for the inhabitant to enable them to use these services effectively. These are also key drivers for BPE in housing.

Rather strangely, the Part L1A regulations state that they are technology neutral and do not require high-efficiency alternative systems or other low and zero carbon systems to be installed. Arguably, accepted technology is never neutral – it simply reflects the normative values of the day. All that is required in these regulations is demonstrable evidence that a study of high-efficiency alternative technologies has been undertaken. However, the regulations disregard such studies if the overall TER and TFEE are met. As such, these regulations are regressive compared to other policy drivers that positively encourage inhabitants to install renewable energy systems as means of reducing their carbon impact. Nevertheless, BPE methodology should always consider the performance of any installed renewable energy systems.

So how well do these building regulations work? One BPE study of over 400 dwellings in England revealed that a third were not compliant with the basic energy efficiency requirements as set out in the building regulations, and that there was a worrying lack of knowledge in the construction sector in relation to these requirements.[31] This goes some way to explaining why housing performs poorly in relation to building regulations, and why regulatory factors on their own can never be enough for developing BPE methodology. Nevertheless, Part L is a strong driver for housing BPE studies, with the TFEE and TER often used as a benchmark for performance.

Healthy homes and neighbourhoods

Health in the home
Health is another key driver for housing BPE, and closely related to our human senses – sight, hearing, touch, taste and smell – as expressions of our wellbeing. Domestic building regulations take account of minimum required daylight levels and acoustic performance in homes, as well as safe water provision, thermal comfort and ventilation.

Typically, homes in the UK are currently expected to be provided with a measured air permeability tighter than or equal to 5 $m^3/(h.sqm)$ at 50 pa in Part F of the English building regulations[32] as well as minimum measured air flow rates in ventilation systems. This is a key health driver and many BPE studies now include air permeability and flow-rate testing within their methods as a means of double-checking compliance. However, to measure the actual levels of toxins present in the home (e.g. polluting gases, particulates and bacteria) requires complex procedures involving sophisticated techniques such as chromatography and mass spectrometry. This is not covered in the building regulations and does not yet form a major driver for most BPE studies, despite the presence of many harmful chemicals in the home (see Figure 2.4).[33]

The high degree of consistency of standards in relation to thermal comfort and allowable temperatures inside homes makes temperature measurement a strong driver for BPE.

Figure 2.4 Toxicity levels in the home should be a key concern of BPE studies

However, people actually experience comfort very differently depending on a wide range of factors, including climate and the degree of control they have over their immediate thermal environment. Some standards, such as ASHRAE 55 and EU EN15251, now calculate different thermal comfort indices with ongoing debate and development in relation to this.[34]

While thermal comfort and indoor air quality are now prioritised by the EU when linking energy efficiency and environmental comfort, the comfort of touch, lighting and acoustics are still seen as secondary factors and are even missing in some EU sustainability tools.[35] As a result, many domestic BPE studies still do not include lighting and acoustic measurements. This is worrying, given that we spend around 90% of our time indoors, and that exposure to excessive noise is considered to be a major and growing health problem in the EU.[36] Nearly half of the population in the UK feel that noise spoils their home life to some extent.[37] The UK standard, BS ISO 177721, introduced in 2018, specifies requirements for temperature, indoor air quality, lighting and acoustics, and provides a common basis for energy performance calculations. It also has a wider application as a basic standard for indoor environmental quality. The UK building regulations Approved Document E: Resistance to the passage of sound also provides a driver for housing BPE studies in relation to sound insulation in homes, which must achieve 45 dB across walls, floors and stairs (see Figure 2.5).

Healthy neighbourhoods, planning and place
BPE drivers mainly relate to buildings but naturally there is always an interaction between a building, its inhabitants, and the immediate environment surrounding the building. There is significant literature on how to set the boundaries for LCA in the built environment sector to take account of toxicity and biodiversity, but relatively little discussion about where the boundaries for BPE should lie, despite some attempts to expand BPE parameters.[38]

Figure 2.5 Noise in homes is a key problem to evaluate in BPE studies

This raises the question: where should BPE studies stop and planning evaluation studies take over? One critical factor is the degree to which inhabitants have pleasant views from their homes as well as access to greenery in the form of either a garden or a local natural amenity nearby. This is because humans thrive when they can see natural settings and enjoy a prospective view.[39] At the same time, the condition of the immediate outdoor environment around a home can significantly affect the degree of comfort within the home.[40]

The government Commission for Architecture and the Built Environment in England broadened the definitions of housing performance evaluation to include qualitative evaluation by inhabitants of their local surroundings, concerning character, access, open space, car parking, safety and security, and local services.[41] This did not develop into a strong BPE driver for political reasons, which precipitated the demise of the organisation and its efforts to promote good housing design.[42] Planning guidance and regulations in some cities, such as London, can now act as drivers for housing BPE to assess the delivery of 'green infrastructure' as built into the housing itself, particularly in terms of 'green walls' or 'green roofs' to replace ground soil lost to the building footprint.

As this chapter has shown, there are a large number of policy and regulatory drivers for housing BPE as well as a number of important potential drivers related to emerging environmental concepts and guidance which have yet to influence BPE. Concerns about climate change and the deployment of diminishing resources will increasingly influence how we evaluate the success of housing using new metrics that compare performance in relation to the *total* resources used. The next chapter develops the theoretical thinking that underpins a new approach to developing housing BPE methodology in terms of the physical factors that set a baseline for the socio-environmental aspects discussed in Chapter 4.

SECTION 2

LEARNING FROM FEEDBACK

THREE
DEVELOPING PHYSICAL
THEORY FOR FEEDBACK

"Give me a place to stand, a lever long enough and a fulcrum, and I can move the Earth."

Archimedes

This chapter explores the role that key physical factors have on housing performance, which must be taken into account in BPE. It frames them in a new way by relating building engineering physics directly to the home environment, and the effects of these factors on energy use, human health and resilience. Understanding the scientific principles and processes that underlie these factors is critical before undertaking any housing BPE study.

The following aspects are covered in this chapter:

- Building science
- Energy use and carbon emissions
- Thermal requirements
- Health in the home
- Housing adaptation for climate change

Building science

Building science is the foundation for evaluating the performance of housing and homes. Without a basic understanding of the key physical principles involved, any BPE evaluator is likely to either misinterpret their physical monitoring data or overlook vital causes in relation to the outcomes. It is a good idea to return to these foundational principles when trying to interpret data or understand the interaction of different interrelated factors of a home as a building system.

Building engineering physics

A key area of knowledge for BPE is the relatively new discipline of building engineering physics, launched by the Royal Academy of Engineering in the UK in 2010.[1] This combines architecture, engineering, human biology and physiology to address energy efficiency, the sustainability of buildings and their internal environment conditions as they affect the inhabitants. It covers the following areas:

- Thermal performance and air movement
- Climate
- Light
- Acoustics
- Construction
- Control of moisture

It also covers the building services supporting the above elements. The overall aim is to use minimal fuel and materials and reduce demands on our natural resources.

Thermal performance and air movement

The two most important foundational physical principles to understand in relation to BPE are the first and second laws of thermodynamics. The first law states that energy can be transformed from one form to another, but cannot be created or destroyed. This relates to heat transfer through one of four processes: conduction, radiation, phase change and convection (see Figure 3.1). The last process is key to understanding air movement and ventilation. The second law states that while the overall energy in a closed system remains the same, the overall amount of usable energy decreases over time during transformation processes, and that heat always flows from hot to cold objects – never the other way around.[2] Some energy is always lost as heat when energy is converted. In simple terms, what this second law means is that for homes to work well there is a need to continually add usable energy to them in the form of maintenance. The concept of 'maintenance free'

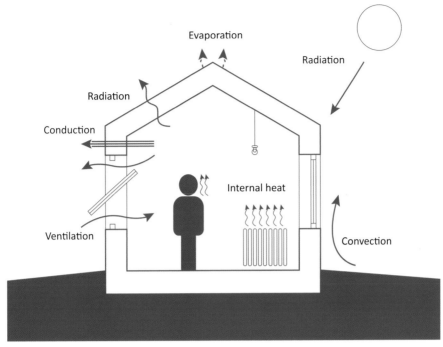

Figure 3.1 Conduction, radiation, phase change and convection all affect housing performance

is a myth – homes always need work to maintain their performance, and to protect against disintegration. Let's look at some of these processes in action.

Conduction

Conduction is the diffusion of heat through a material due to a temperature difference across it, where the conductivity is measured as W/mK.[3] This is important for keeping homes warm, as heat will tend to move through the fabric of the building to the colder space outside. The greater the 'thermal resistivity' of a material as its insulation value, the better it will be at slowing down this heat transfer. The 'U-value' of any material is the reciprocal of the degree of its internal thermal resistivity, combined with its surface heat transfer – the lower the numerical value, the better the insulation quality. The measure of a U-value is Watts per square metre per degree Kelvin (W/m²K), and it is a useful BPE indicator for energy efficiency related to heat demand in homes. Insulation materials also keep homes cool, reducing the rate of heat transfer from a hot outdoor environment into a cooler indoor environment.

Thermal mass

Thermal mass is defined by the specific heat capacity of a material, or its ability to store heat, as well as by its thickness, surface thermal resistance, density and conductivity.[4] The heat flows in and out of a material should align with the changing temperature cycles inside a home. This is so that thermal mass can usefully moderate internal temperatures in relation to the changing temperatures outside within a given period. The moderation works best when there is a large temperature difference between day and night time, but thermal mass can also work successfully in milder climates, providing it is carefully calculated and positioned, and with good night ventilation in place (see Figure 3.2).

Radiation

Radiation is one of the least-understood building physics principles in relation to housing BPE studies. Every material either absorbs or emits thermal radiation, unless it is at absolute zero temperature (0 degrees Kelvin). The radiation from the sun is thermal gain (solar gain), which materials and human bodies absorb. The way in which radiation affects things is determined by their surface properties of emittance, reflectance and absorptance,[5] which change for different wavelengths and different angles of radiation against a surface. This causes a lot of confusion, particularly in relation to the performance of different window systems in different positions in the home. The proportion of solar radiation absorbed through the fabric of a home, or reflected by it, can be decisive in terms of thermal comfort and energy efficiency. However, the immediate 'albedo' (proportion of reflectance of solar radiation) of external surfaces next to the home can also have a major effect.[6] BPE studies often overlook this external factor.

Figure 3.2. The positioning and thickness of thermal mass can moderate temperatures in homes

Convection

Convection is the movement of air or water (or other fluids) through large-scale currents combined with small-scale particle movement from areas of high concentration to areas of relatively low concentration. Warmer, less dense air will tend to expand and rise while cooler, more dense air will tend to fall. Convection can moderate the temperature of a home and remove unwanted moisture and odours either through natural warm air movement (also known as the 'stack effect') or through fans that supply and expel warm air. The 'stack effect' is an important form of natural ventilation for the home, alongside ventilation through differential air pressure between inside and outside due to wind forces. Convection also transfers heat from a solid body to a liquid or gas or vice versa, for example through the exchange of the heat in the air to the surface of a material in a heat-exchanger unit.[7] BPE evaluators need to understand how convection works in homes in order to establish causes of good and poor performance related to moisture movement, heating, cooling and ventilation.

Climate

Housing performance is significantly affected by the natural climate. This can be defined first as the average weather conditions at a place, usually over a period of years, as exhibited by temperature, wind velocity and precipitation; and second as the prevailing set of conditions (temperature and humidity) indoors. Other factors include air pressure, humidity and the degree of sunshine or cloudiness. A thorough housing BPE study will use climate data to contextualise any findings in relation to a home in its local setting. For urban areas, additional factors such as the 'urban heat island' effect also determine the local climate. This is due to excessive solar gain absorption through hard, dark surfaces, air pollution trapping heat, and heat emission from buildings – something rarely taken into account in housing BPE studies, but which is increasingly important.[8] Understanding indoor climate conditions is another critical aspect of BPE and should include an assessment of the human activities such as cooking, washing and window opening which can have a major effect on temperature and humidity levels.

DIF WOODS (CULTURES) - DIF INTERNAL CLIMATES CREATED

Light

The importance of light is underestimated in BPE studies. It consists of the electromagnetic radiation wavelengths from red to violet that cause a visual sensation directly and include white light that is a mix of colours. Light can be created artificially, and it is present naturally due to the Sun's radiation, which gives us all our daylight. The level of light produced is described as illuminance and is measured in lux (lx), where one lux is equal to one lumen per square metre (lm/m2). This describes a quantity of visible light received at any given surface.[9] Designed lighting levels typically deliver around 300 lux for work surfaces. However, the actual level of lighting needed in the home depends on a huge range of human factors, including culture, aesthetics and emotion, as well as comfort, health and functionality. Lower levels of lighting are acceptable for different purposes at different times, and for this reason, local task-based lighting levels are often preferable to prescribed general overall lighting levels in the home. BPE studies tend to focus on how to reduce the energy used in artificial lighting, and comfort factors. Evaluating lighting for health is usually not considered in BPE, though it should be.

Acoustics

Vibrations that transmit through materials and air to reach the human ear cause sound. This deeply affects the human sensory system, as well as building structures. Even subtle background noise can increase stress and the risk of heart disease, while intense vibrations can shatter glass. Sound intensity is measured in decibels (dB) using a logarithmic scale where every 10 dB doubles the loudness perceived. Sound frequency (or 'pitch') is measured in Hertz (Hz). Most humans hear sound levels between 20 Hz and 16,000 Hz, and normally above 0 dB, depending on the sound frequency.[10] Key elements of sound include reverberation (the persistence in time of sound in space after the source has stopped) along with the absorption, transmission and reflection properties of surfaces and materials, which can alter the reverberation as well as the intensity of sound. Sound measurement is often left out of BPE studies, even though it is critical to housing performance (see Figure 3.3).

Construction

The physical home can be loosely be defined as a construction of interrelated parts that is fixed to the ground and is load bearing. Structural integrity defines the degree to which a given structure can hold together under a load, including its own weight, without breaking or deforming excessively during its intended lifespan, for both normal and abnormal conditions. Integrity is important in relation to fire resistance and load-bearing capacity because it establishes whether a home is safe to live in. BPE studies rarely evaluate this because the testing of construction and structures is a complicated and intrusive procedure. Nevertheless it can be relatively easy to roughly determine, using experience and rules of thumb, whether or not the structure of a home appears safe in terms of the visible condition of the structural materials and joints. Durability and the degree of exposure to weathering are also key factors in determining the condition of structures, construction materials and the detailing of the building fabric of the home, which need taking into account.

Control of moisture

All construction materials expand and contract, depending on temperature. This movement can seriously affect housing performance, particularly in relation to the deterioration of weatherproof joints in the exterior fabric of the home. At the same time, materials are subject to the principle of hygroscopicity – the ability to absorb moisture from the air. As a material absorbs moisture it expands, and as it gives up the moisture back to the air, it contracts. This ability of a material to 'breathe' moisture in and out can help to safely regulate the moisture levels in the air within a home, and thus avoid excessive dampness, if all the materials used within a single construction element are 'breathable'.

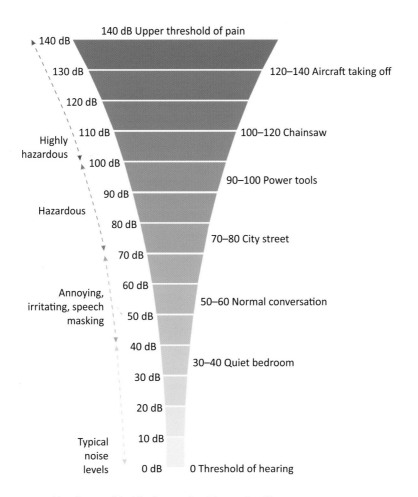

Figure 3.3 Sound levels are critical for human health – scale of impact

When air reaches 100% in terms of its relative humidity, it can no longer 'hold' moisture as vapour, and the moisture starts to condense into liquid water. The most breathable material elements must be situated on the cooler exterior side of the home fabric, and the least breathable on the inside, where it is warmer, to avoid interstitial condensation. This occurs when penetrating warm, moist air inside a wall, floor or roof structure cools down and condenses into water, causing dampness inside the construction. Meanwhile surface condensation occurs when warm, moist air hits a cool material surface and condenses into water. This can be avoided in homes through careful consideration of the thermal insulating and hygroscopic properties of materials. Where surface condensation occurs unavoidably on non-hygroscopic materials, such as in the shower area or kitchen, adequate ventilation and heating is critical to expel the moisture.

Energy use and carbon emissions

Power used in the home

Power is another important principle for BPE. It defines how quickly energy is used or transferred in a given amount of time, and is measured in Watts or Joules per second. Power consumption is critical due to the drastic depletion of the finite natural resources required for power in the home. Even renewable energy requires the use of finite natural resources in the construction of its systems. The key for BPE is to consider how to reduce the energy demand in the home before considering how energy is supplied to it.

Power-hungry 'plug-in' electrical items in the home use lots of energy quickly (e.g. kettles, cookers and showers), or less energy but over a longer period (e.g. alarm systems, digital equipment on standby, and fridges/freezers). These 'unregulated' energy loads are not covered under the building regulations in the UK, but can have a huge effect on how much energy a home uses overall in addition to the regulated loads from heating, ventilation and lighting systems, as well as water use. Evaluating the specification, design, commissioning, use and maintenance of all power systems is critical in determining how much energy they use overall. The assumption here, of course, is that power is always available, though this is not the case in many parts of the world.

Primary versus secondary energy

Understanding the relationship between energy use and greenhouse gas emissions is critical when evaluating housing performance. Energy comprises the primary energy, obtained from natural fuel sources, and secondary energy delivered to a home after it has been artificially processed, such as electricity made from natural gas. Any energy lost during the production, storage, transmission, transportation and delivery stages must be accounted for. Once all these factors have been taken into account, the total energy demand in use of a home can be worked out, with all greenhouse gases emitted as a result calculated as the carbon dioxide emissions equivalence (CO_2e).

Due to the complexity and ambiguity of calculating these emissions, the calculated use of energy is more relevant and robust to work with in housing BPE. However, housing regulation, standards and guidance increasingly refers to carbon dioxide equivalent measurements, given the significance of greenhouse gas emissions in relation to climate change, and BPE needs to refer to these also.

Thermal requirements

Physiological needs

A primary cause of energy use in the home is the need to be able to live within physiologically appropriate thermal conditions. Key factors influencing this are metabolic rate, clothing insulation, air temperature, mean radiant temperature of surfaces, air speed and relative humidity, as well as thermal history. Various socio-technical, cultural and psychological factors can also influence individual perception of thermal comfort including normative values, routines and rituals, expectations, morality, privacy, food, drink and clothing habits and cultural customs.[11] There are two physical methods of calculating perceived thermal comfort – the static model, based on experimental data,[12] and the more recent adaptive model, which takes more account of individual circumstances and differences in reality.[13] The principles of thermal requirements underpinning housing performance are summarised next.

Human metabolism

The human body needs to maintain a relatively stable temperature at its core of 37°C. The human body's metabolism regulates this core temperature through chemical transformations. People have different metabolic rates depending on their activity level and the environmental conditions they experience. The ASHRAE 55-2010 standard defines metabolic rate as the level of transformation of chemical energy into heat and mechanical work, where 1 met = 58.2 W/m^2 (18.4 Btu/h.ft2). This is based on an average person sitting down in the home and doing nothing. It is actually quite a lot of energy, and overall human metabolic energy can contribute significantly towards the heat within a home. BPE needs to account for this factor too.

The human body can physiologically adapt to changing temperature by using food to keep warm and sweating to keep cool, and either increasing or restricting blood flow to the skin surface from its core (see Figure 3.4). Typical behavioural adaptation in the home includes opening and closing windows, adjusting shading, using local fans, and consuming hot or cold food and drinks.

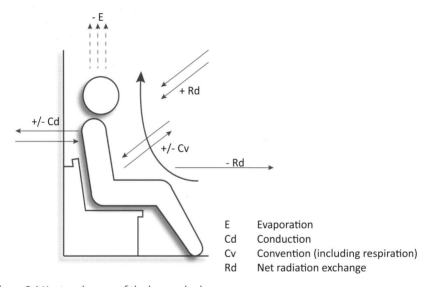

E	Evaporation
Cd	Conduction
Cv	Convention (including respiration)
Rd	Net radiation exchange

Figure 3.4 Heat exchanges of the human body

Thermal environment

The thermal environment surrounding a person includes the following:

- Clothing
- Average air temperature
- Mean radiant temperature (as absorbed by the surface of the body)
- Average air speed
- Relative humidity

All are important factors in BPE diagnostics.

One way the human body cools down in overly hot conditions is through sweating, producing liquid on the skin surface to cool the body as it evaporates. However, in high humidity, it is more difficult for the body to produce sweat that evaporates, and this can lead to dangerous overheating. Recommended relative humidity levels in UK housing are between 30% and 60%, depending on a variety of local factors. The air must not to be too dry, as this can also be uncomfortable.

The thickness, type and number of clothing layers that people wear has a substantial impact on their thermal requirements, too, in terms of reducing heat loss from the body or causing them to feel too cold or too hot. Clothing insulation values are calculated in clo, where 1 clo is equal to 0.155 m^2.K/W. This is equal to wearing a pair of long trousers, a long-sleeved shirt and a jacket.

Homes are generally designed to provide average satisfactory air temperature conditions across all rooms. This approach, however, does not take account of people's thermal sensitivity, which can vary widely, with vulnerable groups such as the elderly, children, the disabled and pregnant women being less able to tolerate temperatures beyond their individual comfort range. BPE needs to consider this variation.

Adaptive thermal comfort

Fieldwork studies using the adaptive thermal comfort model have demonstrated that people will comfortably adapt to a much wider range of temperatures in the home than the original static thermal comfort model proposed by Fanger, particularly in naturally ventilated buildings.[14] There are three categories of thermal adaptation – physiological, behavioural and psychological. This dynamic adaptation depends on the local external temperature and humidity levels people experience outdoors during the course of the year, and the ability they have to moderate their immediate thermal environment, as well as their past thermal experiences and expectations.

In these fieldwork studies inhabitants report their thermal preferences and expectations, and researchers measure the changing thermal environment, in relation to observed behavioural activities, and other psychological/socio-cultural factors influencing inhabitants' expectations and preferences. These studies should form a crucial part of housing BPE studies, but they are expensive to do. Instead, assessment is often reduced to a simple, subjective one-off thermal questionnaire, with no calibration against changes in outdoor temperature over the course of the year, making them less reliable.[15]

Health in the home

A definition of health

Human health is a state of complete physical, mental and social wellbeing and not merely the absence of disease or infirmity, according to the World Health Organization. Physical, mental, social, economic and cultural factors intertwine deeply in this process as we continuously adapt to changes in the home environment in order to stay well. Housing must be designed to accommodate human needs associated with health and wellbeing given that, on average, people spend two-thirds of their time inside the home in EU countries.[16]

Indoor air quality

Indoor air quality (IAQ) is increasingly important for BPE, as homes become more airtight and trap accumulative indoor pollution for longer periods.[17]

Key indoor contaminants include the following:[18]

- Carbon dioxide (CO_2) and other human bioeffluents
- Gaseous volatile organic compounds (VOCs)
- Tobacco smoke
- Carbon monoxide (CO)
- Radon
- Nitrogen dioxide (NO_2)
- Bacteria, fungal spores, mites
- Fibres
- External pollution such as particulates

The average adult breathes in 13,000 litres of air a day, and breathes out carbon dioxide into the home. High levels of CO_2 are associated with stuffiness and poor concentration. The warm water vapour produced by people and their activities can also lead to the growth of harmful fungal spores in the form of mould and mites in the home, which have been associated with asthma.[19] Formaldehyde as a VOC is a known human carcinogen, causing sensory irritation at low concentrations. CO_2 is the leading cause of death from indoor chemical pollution due to faulty equipment and tobacco smoke.[20] Low levels of chemicals present in the home can produce additional health problems, which when combined are even more dangerous for vulnerable groups such as young children and the elderly.[21] Although CO_2 levels are frequently seen as a good proxy for general indoor pollution, there is not necessarily a correlation between CO_2 and other chemical levels.[22]

Daylight and sunlight
Diurnal and seasonal light variations can have significant health impacts, as can the quality of light in terms of colour and spectrum. Daylight is vital for health with regard to regulating the human body's production of the hormone melatonin, which governs sleep patterns, as well stimulating the body's production of serotonin, which can reduce symptoms of depression.[23] Sunlight in the form of radiation has further benefits for wellbeing, including warmth and, interestingly, disinfectant properties which kill bacteria. Many countries, including the UK, still do not provide adequate regulations covering how much daylight or sunlight there should be in homes. Guidelines tend to refer to minimum 'daylight factor' levels for rooms. But this calculation does not take account of visual comfort factors such as glare, and takes no account of the location of a home. Further EU guidance is being developed in this area.

Noise
The impact of noise on human health in the home depends primarily on its volume, duration, degree of repetition and frequency. Typically levels of speech at around 65 dB can be reduced through good sound insulation in walls to about 20 dB (barely audible), although humans tend to find high- and low-pitched sounds more disturbing than those at middle frequencies. Even normal background noise levels can cause disturbance cumulatively. The complete absence of sound can be equally unsettling. The degree to which people tolerate noise also varies widely. Excessive noise in the home can lead to annoyance, anger, frustration and sleep disturbance, with associated startle and defence reactions as well as speech interference. These can in turn lead to deeper physiological harm such as eardrum damage or hearing difficulties, and cardiovascular effects with increased blood pressure, brought on by increased stress levels, as well as psychological damage.

For these reasons, evaluating noise in homes is very complex and goes well beyond simply assessing the degree of sound insulation present, although this is a good place to start in BPE studies.

Radiation

Harmful radiation can be present in the home through the following:

- Solar ultraviolet radiation
- Electromagnetic fields (EMFs)
- Radon gas

Ultraviolet radiation does not pose a great threat to inhabitants, as glass blocks most of this radiation. There have, however, been leukaemia clusters associated with housing within close range of EMFs emitted from overhead transmission cables,[24] and concerns raised about the effect of this type of radiation combined with other health-affecting factors.[25] BPE needs to take EMFs more seriously, but the effects are hard to isolate and measure compared to measuring radon and solar radiation. The presence of radon gas in the home is typically due to a build-up from naturally occurring sources in the ground (e.g. granite), which occurs in many regions of the world. Preventative measures are regulated for in many countries, as it has been shown to have a significant link with cancer.[26]

Housing adaptation for climate change

Evaluating adaptive capacity

Designers and BPE evaluators need to consider the capability of a home to be retrofitted to adapt to future climate change.[27] One way forward is to build in future weather patterns at the design stage and 'stress test' the design of homes in relation to these factors through advanced modelling techniques such as computational fluid dynamics associated with building information modelling.[28] Evaluators can also retrospectively feedback on the performance of the home during extreme weather events and provide recommendations for any future upgrading required. Finally, it is possible to assess the degree to which key climate adaptation strategies have been adopted in a home.

Overheating

BPE needs to evaluate the degree to which unwanted heat is kept out of the home in the first place, before ventilating it away, through the provision of user controlled external solar shading to prevent solar heat gain, increased external surface reflectance (e.g. painting surfaces white) and extra insulation as an optimised package.[29] Windows should open inwards to maximise opportunity for adding external shading devices in the future, such as shutters and blinds (see Figure 3.5).

Flooding

Evaluating adaptation for future increased flood risk includes assessing the ability to slot in vertical waterproof plates across the bottom of door openings, or flood-proof doors, as well as ensuring the provision of waterproof ground floors, blockable water drains, and waterproof lower wall areas in the event of rising water levels.[30] Perhaps the most vital requirement is for means of access to the roof from single-storey dwellings, to prevent possible drowning – rarely thought of, and rarely evaluated in BPE.

Figure 3.5 Evaluating the ability of homes to adapt to climate change should be a key part of BPE – external shading helps to avoid overheating

‣ Storms

Hurricane-force winds are increasingly common, and monsoon-like conditions are spreading to many parts of the Northern Hemisphere, with super-intense downpours that can last for days. The term 'weatherbomb' has now entered common usage to describe these phenomena. Adaptive measures to evaluate here include the inclusion of more robust and larger fixtures and fittings in relation to roofing and rainwater goods, as well as increasing the porosity of any areas adjoining the home (e.g. permeable pavements) to help absorb and slow down the rain run-off.

Drought, energy and water resilience

Finally, the loss of power and water supplies through storms, heatwaves and drought is a real and present danger for homes in many countries. Adaptive measures to evaluate here include the potential to add on back-up power supplies such as off-grid renewable energy systems (solar, wind or hydro), or store back-up fuel, as well as the ability to add rainwater storage systems which can help with extra demand over the year.

Safety

Safety is a key physical performance factor in housing, discussed earlier in relation to structural loading. However, safety considerations in the home are far wider than this, and include an evaluation of safe access, use, maintenance and replacement of all key services, as well as all construction. Despite safety regulations, it is surprising how often windows are left relatively inaccessible for cleaning in new homes, and how difficult it can be to reach equipment and controls.[31]

Having set out the key physical factors to consider in housing BPE, the socio-cultural context in which they operate is examined next to provide a more effective overall approach to BPE.

FOUR
DEVELOPING SOCIO-CULTURAL THEORY FOR FEEDBACK

" ... conventions that are often taken to constitute the context of behaviour have no separate existence: rather, they are themselves sustained and changed through the ongoing reproduction of social practice "

Elizabeth Shove in 'Beyond the ABC: Climate change policy and theories of social change', Environment and Planning A, June 2010, vol. 42, no. 6, p 1279

Homes that are identical in design and construction can still use very different amounts of energy – up to 14 times difference in some cases[1] and, as Janda has eloquently pointed out, '... buildings don't use energy, people do'.[2] Understanding the relationship between physical and socio-cultural factors in housing performance is therefore critical for housing BPE.

This chapter explores the wider role of theory underlying housing performance evaluation, with specific reference to the following:

- The socio-cultural focus
- Bringing the disciplines together
- Needs
- Perception and information
- Behaviour, negotiation and practices
- Motivations
- Participatory planning, design and learning

The socio-cultural focus

The study of the interrelationship between technology and people in their homes has benefited significantly from insights provided by human geography,[3] psychology,[4] sociology,[5] human factors[6] and ethnography,[7] among other disciplines. Methodologies coming out of this work are increasingly applied in various housing BPE studies and are important to consider before deciding on the choice and use of various BPE methods and techniques. Socio-technical studies also examine the broader societal aspects shaping the interactions that go on in the home.[8] These aspects form an important context for BPE studies, linking housing performance to deep structural issues such as governmental framing of context and subsequent policy and funding.[9]

One aspect consistently marginalised is how history and culture affect housing performance.[10] Socio-technical studies tend to be more concerned with what is happening currently in society:[11]

'Whilst understanding the longer sweep of historical change is important ... focusing on more recent and on-going dynamics of change taps more directly into policy concerns for understanding how energy demand is changing "now" and is likely to evolve into the future.'

By contrast, cultural studies demonstrate how the deep past continually affects specific groups in terms of beliefs, values, rules of behaviour, habits, expectations, and the evolutionary processes within which these change over time. For example, as the demand for more heating in the home evolved in the UK from the 1920s onwards, this led to various historic practices which still affect thermal habits today.[12] Another example is how inhabitants in different cultures react to the idea of opening a window in a home, according to climate, social norms, custom and habit.[13] Sleeping with a window open at night is a typical cultural behaviour in some countries.[14] It takes time and concerted effort to change ingrained home habits.

For BPE to be effective it must consider the history and culture of any home environment and its inhabitants, as well understanding how contemporary BPE methodology itself is influenced by these factors.[15]

The term 'socio-cultural' is used here to highlight the interrelationship between social and cultural aspects in housing BPE as they impact on evaluations of the home environment and its technology. All are deeply interrelated with the building science principles described in the previous chapter (see Figure 4.1).

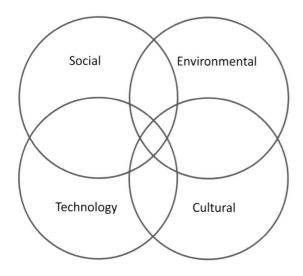

Figure 4.1 BPE is impacted by the interrelationship between the socio-cultural axis and the environment–technology axis

Bringing the disciplines together

Housing BPE is informed by many different disciplines. How, then, can these to be brought together to establish a better understanding of housing performance and generate new insights? In practice, delivering a genuine socio-technical study can be challenged sometimes by incommensurable research design and subsequent data which cannot be compared between disciplines.[16]

Knowledge generated through the collaboration of disciplines in BPE can be considered as the following:

- Multidisciplinary – collaborating but using different paradigms and separate work
- Interdisciplinary – working more closely together with an agreed shared language
- Transdisciplinary – working together very closely to generate new theory and language[17]

It is unusual for engineers and architects to work directly with social sciences and/or humanities academics on housing BPE studies, due to the difficulties of transcending the different paradigms that each discipline sits within,[18] as well as narrow educational curricula stuck in disciplinary silos. BPE studies typically engage at a multidisciplinary level when the engineers/physical scientists do the physical monitoring of the home, and the architects/social scientists carry out a separate package of work related to the inhabitants.[19] Engineers and social scientists also carry out studies related to housing performance without engaging with each other at all.[20] There are, however, some excellent interdisciplinary projects studying the relationship between inhabitants and technology[21] and combining health and engineering insights.[22]

Recently, engineers working with social scientists and physicists have developed a transdisciplinary understanding of predicted building performance using 'agent based modelling' to create relatively predictable inhabitant behaviour within a dynamic physical model.[23][24] This type of transdisciplinary integration holds much promise for housing BPE providing it fully engages with inhabitants' needs.

Needs

Needs theory

Needs arise '… when there is a discrepancy between the observed state of affairs and a desirable or acceptable state of affairs'.[25] Inhabitants' needs are defined in terms of what the human body and mind require to function well. These definitions are then incorporated into government policy and standards for housing, but the value systems underlying such definitions often remain tacit. Socio-cultural theories related to BPE question these value systems in order to define what inhabitants' comfort needs actually are.[26] Recent research on assumed thermal comfort requirements has helped to explain how these tacit normative values are embedded in regulations.[27] Needs, however, constantly change as they emerge and disappear through inhabitants' engagement with new technologies.[28]

Despite criticism, Abraham Maslow's Hierarchy of Needs[29] is still a useful definition of human motivation in terms of biological and physiological needs – safety, love and belongingness, esteem and, finally, self-actualisation[30] (see Figure 4.2).

Manfred Max-Neef expanded these needs into a broader non-hierachical list:[31]

- Subsistence
- Protection
- Affection
- Understanding
- Participation
- Leisure
- Creation
- Identity
- Freedom

All of these needs clearly relate to housing, and yet most BPE studies tend to focus on only energy, water, building fabric, comfort and health needs.

The 17 Sustainable Development Goals[32] promoted by the United Nations in 2015 are the latest global definition of human needs evolving from Max-Neef's original work. Key goals related directly to housing BPE include the following:

- Good health and wellbeing
- Clean water and sanitation
- Affordable and clean energy
- Sustainable cities and communities
- Sustainable consumption and production
- Climate action
- Life on land (biodiversity)

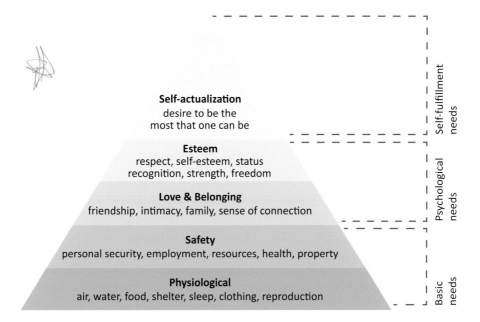

Figure 4.2 Maslow's Hierarchy of Needs is still a useful way of framing BPE concerns

While not a fully comprehensive list, this is probably a good place to start in terms of seven basic areas that should be covered by housing BPE methodology.

Affordance theory

For these needs to be met, inhabitants need to be able to use their homes effectively. This should happen intuitively, so that it is 'obvious' how to live in a designed home. Affordance theory[33] suggests that the 'persisting surfaces of the environment … provide the framework of reality'. Affordances are thus described by Tim Ingold as 'properties of the real environment as directly perceived by an agent in a context of practical action'[34]. A chair affords us the act of sitting – we intuitively know to 'sit' in the chair. Importantly, however, an observer may or may not perceive the available affordance, depending on their needs and background.

Affordance theory says that things which look like what they actually are provide appropriate information for perception, whereas things which don't look like what they actually are provide dangerous environmental misinformation for the perceiver. Furthermore: 'the perceived meaning of the object does not reside solely with the object or the person, but it derives from the intentional relationship between the two, which includes the history of their interactions'.[35] Affordance theory deeply informs user-centred design in relation to product development,[36] but it has not really been applied in housing BPE methodology, beyond the odd usability case study.[37]

Usability

Three critical principles for effective user-centred design[38] help ensure a home that is fit for purpose:

1. Early focus on users and tasks
2. Empirical measurement – testing prototypes
3. Iterative design as a result of 1 and 2

So why are usability studies neglected during the design and BPE process in housing? Often inhabitant diversity is underestimated, or it is overestimated and thought not possible to study the diverse usability. Architects assume they know how homes are used based on their own experience, even though this ignores other people's understanding. There may be a pre-determined belief in the effectiveness of technology, or in the idea that design iteration is an overly expensive fine-tuning exercise which delays the contract, and that guidelines should be sufficient for the user to understand how to use things.[39] In fact, the pre-testing of housing prototypes for usability can be hugely beneficial both in retrofit[40] and newbuild homes.[41] This can provide significant insights into how inhabitants directly experience technologies in the home in terms of usability, and the challenges they face.

The really worrying thing is that people usually adapt well to a required task in their home, despite an inefficiently designed tool for the task.[42] The adaptation process is not understood precisely because inhabitants are so good at creating workarounds. The problem of usability is thus masked, because inhabitants have to tolerate the original design faults. Just think of how many wedges you have seen jammed into poorly operating doors and windows, as an example of this (see Figure 4.3).

Perception and information

Cognitive perception

Understanding how inhabitants perceive their home environment is critical for housing BPE. This establishes a link with performance issues and identifies the underlying reasons for the mismatch between design intentions and actual use. There are different and contested information theories related to human perception.

For cognitive psychologists, perception depends on prior experience, which provides a mental 'mapping process'[43] through which the 'landscape' of the home is understood. The perceiver actively chooses what they see, selecting material objects for attention and perceiving some of their properties rather than others, as filtered by the perceiver's cognitive schemas. Importantly, the inhabitant can take in new information and adjust to it, changing the 'map' for future reference. Perception is thus seen as a constructive process based on experiences and provides a basis for memory.

Ecological psychologists tend to refute cognitive perception theory and follow the cogito-sensory idea of affordances.[44] This is closely associated with phenomenology as a philosophy which states that people understand the world primarily through sensorial experiences.[45] A phenomenological approach to housing BPE is primarily subjective and offers no means of creating a predictive model for understanding housing performance, but it can offer powerful insights into how and why people actually experience buildings in the way that they do.[46] This can be very useful to designers, facility managers and clients.

Information theory also draws on language and symbolism, not as aspects that *mediate* perception but as aids which can actually help to *direct* perception.[47] These aspects need to be carefully evaluated in BPE, particularly in relation to the many environmental controls in the home, which depend so heavily on the language and symbols of guidance when the intuitive affordances are not well designed[48] (see Figure 4.4).

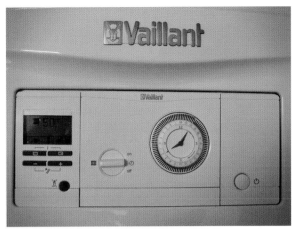

Figure 4.3. A wedged door is a good indication that a home is not performing well

Figure 4.4 Language and symbols provide a key means of navigating home use and should be evaluated in BPE

Behaviour, negotiation and practices

Planned versus negotiated behaviour
Perception is just one component of evaluating how people interact with their homes. Several socio-cultural approaches are currently used in housing BPE studies to develop this understanding further in terms of 'how' and 'why'.

Behavioural theories related to housing performance stem largely from psychology. Planned behaviour theory assumes that inhabitants' decision-making and subsequent behaviour is planned to achieve logical goals based on norms, expectations and perceived control.[49] This approach has been adopted in various housing BPE studies[50][51] to understand the behavioural causes underlying housing performance. One of the key criticisms of behavioural theory, however, is the separation of people from the influence of home technologies or energy policies.

Negotiated behaviour draws on Latour's Actor Network Theory of power relations,[52] which basically states that a home consists of a network of human and non-human 'actors' who either act passively to preserve the status quo or actively with the power to create changes in housing.[53]

As Grandclément et al. put it:
'... energy performance is not fixed by the design team but is continuously negotiated. The alignment of the technological system and the user is a process that continuously unfolds in multiple arrangements with different stakeholders and varying interests. Instead of assuming the static, techno-material quality of a building, this concept invites us to consider energy performance as a more fluid socio-technical process to be performed throughout the life of the building and its inhabitants.'[54]

Latour also gives an example of how technical objects have power over humans based on the assumptions 'inscripted' into the object through its design and making.[55] When an automatic door closer fails to function according to the design 'script' associated with it, it forces humans to try and carry out the work themselves, even if this is not optimal and is resisted by the technology. Indeed, Latour goes on to say 'If, in our societies, there are thousands of such lieutenants to which we have delegated competences, it means that what defines our social relations is, for the most part, prescribed back to us by nonhumans.'[56] In other words, it is construction and technologies which also define how inhabitants live and relate to each other in their homes, because designers give them the power to do so. BPE needs to focus upstream of inhabitant behaviour, and understand how technologies influence this.

Practice theory

Behaviour and BPE studies often ignore the meanings, habits and choices of individuals and groups as they are shaped by cultural beliefs and the societal forces that create and define housing needs.[57] Social practice theory, as set out by Theodore Schatzski, addresses this deficit.[58] He extends Latour's Actor Network Theory of power relationships to show how people's 'sayings, meanings and doings' are 'bundled up' together with other societal influences in relation to something material, to create a 'practice'.[59] One example is the 'simple' programmable thermostat, which is:

'... simultaneously influenced by the user's desires ... and by "affordances" of the material design of the thermostat ... which also incorporates assumptions about the user ...'[60]

Housing 'practices' can be understood more closely in terms of four key elements in housing:

- Technologies and products
- Institutionalised knowledge and explicit rules
- Inhabitants' know-how and habits
- Engagements which include purposes and tasks [61]

The interrelationship of these four elements cannot be disaggregated, and requires particular BPE methods of enquiry and analysis to ensure that the extent of the 'bundled' practice is considered across each of them.

Motivations

Unpacking the practices taking place in the home reveals other important factors, including inhabitants' expectations, aspirations and emotions, which inform their motivations. As with many other socio-cultural factors discussed in this chapter, these tend to be largely ignored in housing BPE methodology, but they shouldn't be.

Expectations

People's values tend to drive their core beliefs, which in turn can affect how they interact with their homes. Personal Construct Theory[62] states that people create their constructs of the world through testing them experientially. The deeper the construct, the harder it is to dislodge. Typically, core constructs relate to core needs. An inhabitant's 'constructs' form the historical basis of their expectations in the home, derived from their own life events. These expectations are not always logical, but they are very real to the person concerned. If they have lived in a cold house before, then a new home may feel uncomfortably warm to begin with, until they adjust their expectations. These thermal expectations can also be heavily influenced by what is promoted as the 'new normal' by society, government policy and technology development. When central heating first arrived, people demanded more heat in response. When TV first arrived, their newly sedentary lifestyle also demanded more heat – they expected to be able to sit still for long periods of time.[63] BPE therefore needs to understand people's expectations.

Aspirations

Aspirational values in the home can be quite different from expectations, however. Aspirations relate to inhabitants' cultural tastes and status seeking. They include aesthetics,[64] heritage,[65] wanting to carry out DIY in order to feel more at home, making a 'dream kitchen or bathroom' even when this is not functionally necessary, and wanting something new and fashionable,[66] as well as having environmental aspirations.[67] Sometimes the aesthetics and aspirations of designers and inhabitants get in the way of using technology effectively. This was the case with inhabitants in the LILAC co-housing community in Leeds, UK, where their photovoltaic inverter meters were deliberately placed out of sight, reducing their interaction with them. Inhabitants' aspirations have been grouped along with other attitudes and behaviours into five generic human 'personas' through which to understand how homes should be designed and improved for diversity.[68] These range from 'the Stalled' and 'the Pragmatist' to 'the Property Ladder Climber', 'the Affluent Service Seeker' and all the way through to the 'the Idealist Restorer'. How is housing BPE methodology supposed to deal with understanding inhabitants' aspirations? By us talking with them and observing what they do.

Motivation

Motivation drives an inhabitant's behaviour in the home and can explain why a home is not performing as expected by the designer. Although motivation is traditionally associated with logical planned behaviour in housing BPE studies,[69] it can be easily driven by relatively 'irrational' aspirations when it comes to fashion, aesthetics and prior experience, which still make sense to the inhabitant. It can also be reduced by a lack of understanding. This happened when the LILAC co-housing inhabitants in Leeds felt no need to examine options for reducing their use of electricity from the grid. They did not consider replacing it with renewable energy from their photovoltaic systems, because they erroneously believed their energy use was already low.[70]

Environmental motivation is an assumption made in many government policies, designed to encourage user behaviour change in relation to reducing energy use in the home, based on rational decision-making. However, inhabitants may claim to have environmental values while failing to take action for a variety of complex reasons.[71] Financial savings can motivate inhabitants in relation to improving their homes to save energy, but this can also be overridden by 'irrational' motivations prompted by the attitudes and aspirations described above.[72] Housing BPE also needs to take account of unforeseen 'pre-bound effects', where people end up saving less energy than they were supposed to, because they were motivated not to use it in the first place due to their financial constraints. This is not accounted for in standard energy use predictions.[73]

Emotions

Emotions are rarely discussed in housing BPE studies, and yet they can have a very powerful influence in terms of how an inhabitant is motivated to engage with their home. The design of a home can arouse strong emotional attachment, and it can be 'cherished' by its inhabitant as a result. Much of this is to do with how much empathy the home evokes in the inhabitant and how much it offers itself to be looked after.[74] A so-called 'maintenance-free' home can actually disengage the inhabitant. This can have a major impact in terms of

NEEDS TO NEED CARE/ MAINTENANCE

how well a home is maintained and repaired, which in turn has consequences for housing performance. Emotions are conditioned by a variety of physical, mental, social and cultural factors, and are often complex and contradictory in nature – which may explain why they have not been taken into account in BPE methodology to date. But this does not mean they should be ignored, as they may well offer explanations for poor performance either through causative or consequential effects.

In summary, it is vital for housing BPE to take account of all of the dimensions above, which help with understanding an inhabitant's practices in the home. These dimensions can be summarised in the following diagram (see Figure 4.5) for an overall theoretical approach to housing BPE.

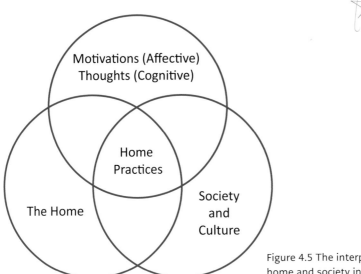

Figure 4.5 The interplay of inhabitant, home and society in BPE

Participatory planning, design and learning

Linking BPE with planning factors

Planning policies can determine the siting, density and layout of housing, and its allowable form and materiality, as well as the co-location with other types of buildings, transportation, water and power services, and green infrastructure – all of which affects how well homes perform in relation to inhabitants' needs and wider sustainability. If a BPE study does not take these planning factors into account in terms of context, it will probably overlook obvious reasons why the indoor air quality is poor (nearby factory or overly dense development?), or why the energy used for heating is excessive (no sheltering buildings or greenery?), for example.

Conversely, BPE findings rarely feed back into spatial planning policy development, although there are now calls for a greater understanding of the dynamics of the wider unintended consequences arising from energy policy interventions in buildings.[75] Equally, numerous international BPE studies have shown that significant overheating is occurring in housing,[76] but relatively few have examined the external spatial planning issues that affect indoor overheating in homes.[77][78]

Interactive adaptivity

If housing BPE is to be truly effective in influencing future housing design, it needs to be able to evaluate the performance of the home more holistically. This means taking additional criteria into account as part of its developing methodology. In addition, the barriers between physical and socio-cultural aspects of BPE need to be removed. To this end, Raymond Cole and others have developed a new approach to understanding the relationship between people and buildings, called 'interactive adaptivity',[79] explaining that:

'A dynamic and complex building system with a participatory process, interactivity between inhabitants, and between inhabitants and building elements can adapt to changing conditions (e.g., seasonal temperature change, or, on a larger scale, global climate change), resulting in a fluid but robust design that is responsive to social, ecological, and economic conditions over time.'

This idea of 'interactive adaptivity' is important in housing BPE studies when evaluating the opportunities inhabitants have to actively manage the environmental conditions of their home via the physical building and services,[80] and how the home in turn learns about its inhabitants' needs. Some people claim that 'smart homes' are a good example of this type of interaction, but 'smart' controls can also take away people's power to control their environments.[81]

Collective learning

Understanding how inhabitants learn *collectively* to interact with their homes and neighbourhood is another exciting new methodological approach for housing BPE, which has traditionally only concerned itself with how individuals or individual households interact with their home. Social learning theory can be used in BPE studies to explore how inhabitants help each other to learn how to use their homes more effectively, either through groups such as co-housing communities[82] or through emerging social media communication between inhabitants in an individual housing development.[83]

The approaches to BPE outlined in this chapter need to be considered collectively for a deeper understanding of housing performance. They offer a new way to think about traditional housing BPE in relation to design modelling and evaluation, as set out next.

FIVE
MODELLING AND REALITY

"What gets us into trouble is not what we don't know, but what we know for sure that just ain't so."

Mark Twain

Having set out a theoretical background to housing BPE, this chapter describes how 'real world' action research[1] can feed back into the housing design and build process. This is essential in order to improve the accuracy of models developed for predicting building performance. There is often an over-reliance within the design team on digital modelling processes at the expense of testing reality. Modelling assumptions are questioned and examined more closely here. The significance of commissioning and behaviour is highlighted alongside the crucial question of 'how much modelling and feedback is enough?', which is tackled in relation to the various BPE methods available.

This chapter covers the following aspects:
- Predicted performance
- Performance in reality
- Commissioning the home
- Inhabitant engagement with the home
- How much modelling and feedback to do

Predicted performance

Modelling and housing BPE

In housing BPE, a model is a simplified description of a real system or process to assist calculations and predictions of housing performance. It can be either conceptual, representing a system using rules and concepts (including socio-cultural aspects) or it can be scientific, offering a simplified and idealised understanding of physical systems, as set out in Chapter 3. Conceptual models are often abstract generalisations of things in the real world, whether physical or socio-cultural (see Figure 5.1).

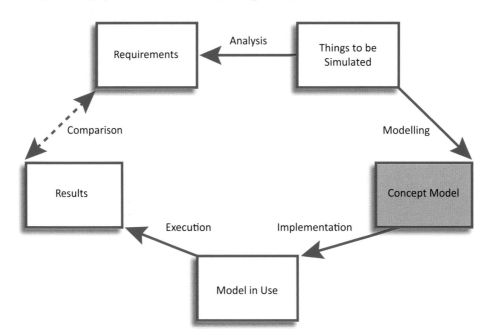

Figure 5.1 A conceptual model in relation to the design process

A model simplifies and simulates what happens in the home. It can describe how a home ought to perform according to design intentions, by defining the fundamental principles and basic functionality of the home 'system' which it represents. A steady-state simulation only covers an instant in time, whereas dynamic simulation provides information over time. Both can be very useful for testing, analysing and improving housing design, providing the model is accurate enough.

Validity is the degree to which the model faithfully represents its system counterpart in reality. Zeigler characterises three types of model validity:[2]

- Replicative – i.e. whether the model fits the data already acquired from a real system
- Predictive – i.e. whether the model fits the data before they are acquired from a real system
- Structural – i.e. whether the model completely reflects the way in which the real system operates

Validity is important for BPE, which relies on accurate design models to compare against housing performance.

Using system theory for housing BPE

A system is a bounded entity with interrelated and interdependent parts, with the complex whole being greater than the sum of its parts. Changing one part of the system affects other parts and the whole system, with predictable patterns of behaviour. Systems describe and predict physical reality, as well as interactions between social and physical aspects. The models are refined by testing them against real situations. Any system makes use of inputs and outputs in order to do purposeful work. Feedback loops help the system self-correct and maintain itself (see Figure 5.2). Open, living systems have emergent properties and tend to evolve, whereas closed mechanical systems generally have fixed properties, and these properties only change with human input or a technical malfunction.

In this sense, a home is a modelled fixed thermal system, for example, where:

Q_i = internal heat gain
Q_c = conduction heat gain or loss
Q_s = solar heat gain
Q_v = ventilation heat gain or loss
Q_e = evaporative heat loss
using the equation: $Q_i + Q_c + Q_s + Q_v + Q = \Delta S$, where ΔS is a change in heat stored in the house.

The house is thermally balanced when ΔS is zero. Otherwise, the house is getting either warmer or cooler, according to the above inputs and outputs. A typical mechanical thermostatic control unit as part of a heating system provides one feedback loop for these inputs and outputs. It can provide 'negative' feedback, which shuts down internal heat gain (the heating system) if too much solar gain input is putting the system out of balance. Alternatively, it can provide 'positive' feedback that opens up the heating system, if the temperature in the system is too low. The feedback depends on the temperature setting of the thermostat.

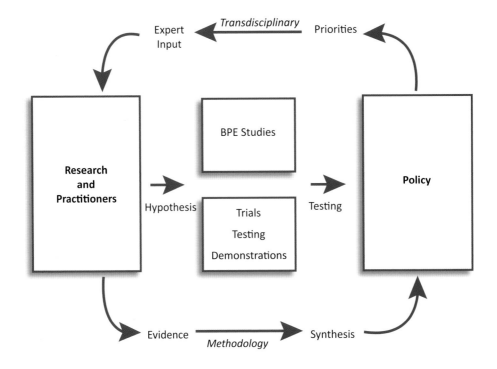

Figure 5.2 BPE is part of an epidemiological system which provides feedback to correct itself

A typical home environment system can thus either be considered in this way under non-changing conditions as in steady-state modelling, or it can be considered under conditions which change over time using dynamic modelling.[3] Housing BPE tends to use both, with different BPE methods for testing and validating the modelling in question.

Steady-state models as benchmarks for BPE

The BRE Domestic Energy Model (BREDEM), developed in 1992 as a simplified steady-state model, underpins building regulatory standards in the UK. It calculates the predicted energy use and fuel requirements of homes based on their physical characteristics, using inputs from the fixed thermal system illustrated above, adding in customised values for energy consumption related to space and water heating and cooling, lighting, cooking and appliances, and offsetting this consumption against any onsite renewable energy generation.

The Standard Assessment Procedure (SAP 2012) is used by designers to comply with the UK building regulations. It derives from BREDEM and includes generic assumptions about how residents will use heating (thermostat settings and hours of use), lighting and hot water. It also estimates energy and carbon savings from low- and zero-carbon technologies. An even more simplified version of SAP, known as RdSAP, calculates energy use in existing homes where it may not be possible to determine the construction and servicing of the home. There has been much criticism relating to the effectiveness and accuracy of SAP and RdSAP assessments due to their over-simplification.[4] It is difficult to use BPE to demonstrate the degree to which a home has met the design standards intended, when the predictive model is weak in the first place.

A more sophisticated steady-state model generated by the Passivhaus Planning Package (PHPP), first developed in Germany in the early 1990s, includes inputs to take account of thermal gains and losses, particularly in relation to thermal insulation, thermal bridging, airtightness, ventilation and solar gain. This model is a design tool rather than just a compliance tool, and it is based on a series of Excel spreadsheets.[5] On average, BPE studies of Passivhaus-standard homes have generated results that are more or less in line with PHPP predicted energy use overall, suggesting that this model is robust in reality.[6]

Dynamic modelling

Dynamic modelling of predicted housing performance requires complex calculations to account for all the variables over time. Typical commercially available computer programmes for use by the design team include EnergyPlus, IES, EcoDesigner and DesignBuilder, as well as TRNSYS, with variable capabilities that have been compared.[7] These programmes are able to accommodate computer-aided design (CAD) drawings with associated lighting, thermal, acoustic, ventilation, and life-cycle costing simulation packages. However they still do not reflect the reality of housing performance, as they are based on an approximation from averaged default values, and are unable to take accurate account of inhabitants' behaviour.[8] More sophisticated models of building performance are needed to take account of externalities beyond the home as well as more complex interrelated factors, showing how the construction of homes responds to energy use and humidity, and the implications for human health.[9] These programmes also need to evaluate homes on a room-by-room basis, acknowledging the different sensory and affordance qualities in each space, to represent housing performance in its totality.[10]

Building Information Modelling and BPE

Building Information Modelling (BIM) is a new horizon for housing BPE, allowing all building information to be fully coordinated within a single computer-generated building model. Individual 'objects' of information are generated which contain both the dimensions and the specification of any component, as well as any other relevant data. These 'objects' are then coordinated within the three-dimensional space of the model. This should improve building performance, with all design team members working together on exactly the same model. There is also an excellent opportunity to introduce BPE feedback directly into BIM packages via the facility management coordination packages that could allow real-time monitoring, potentially enabling housing managers and inhabitants to fine-tune the performance of their homes. But this has yet to materialise, as most housing developers are not ready for BIM,[11] and there is no readily adopted BIM strategy for housing maintenance and refurbishment in the UK.[12]

Performance in reality

The gap between predicted and real performance

To understand exactly how deep the performance gap really is between predicted modelling and reality, it is worth reflecting on the recent meta-analysis of 76 homes from 59 projects in a national UK £8-million BPE programme,[13] which revealed the following:

- Average total carbon emissions were 2.6 times higher than the average design estimate.
- None of the 'zero-carbon' design estimates were achieved in practice.
- There was scarcely any link between the SAP estimate of CO_2 emissions for space and water, heating, lighting and ventilation, and the actual total emissions.
- In nine out of thirteen cases, the actual measured heat-loss co-efficient (whole building fabric insulation value) was worse than the SAP design estimate.

- Key areas of concern were poor implementation of new technologies, unnecessarily complex controls, inadequate training of installers, poor commissioning and handover procedures, and poor fabric construction, resulting in extensive thermal heat loss.

Another recent meta-analysis of 60 homes across 19 projects within this same BPE programme revealed significant performance gaps, with 57% of bedrooms and 75% of living rooms suffering from summer overheating based on the Chartered Institution of Building Services Engineers (CIBSE) threshold of 5% annual occupied hours at more than 25°C.[14] Indoor air quality is another major performance gap in housing. A study of eight newbuild low-energy affordable homes in the UK revealed that in winter, carbon dioxide levels reached levels as high as 3427 ppm in the living room and 4456 ppm in the main bedroom, or close to a health risk threshold.[15]

The persistence of the performance gap

This major performance gap persists even though there has been plenty of international research over the decades as well as the introduction of new technologies and modelling systems. Therefore, there must be deeper causes that prevent the obvious 'solutions' from taking hold.

Sometimes the deeper cause is due to the design team and researchers attributing the findings from housing BPE studies to the category of 'human behaviour'. They wrongly blame the occupant or believe that the answer lies in 'educating the occupant' as stated in a UK National Energy Foundation report[16] when, in fact, the cause lies blatantly in the poor design and installation of technology. This makes it impossible for the hapless inhabitant to use it properly[17] (see Figure 5.3).

At other times, the fault lies in a blind faith in technology with a refusal to believe that it can go wrong, and a belief that the inhabitants must be 'interfering' with it.[18] This often occurs when engineers believe they understand better how housing performs than social scientists. This is a mistake, and many housing BPE studies engage in 'bean counting' engineering

Figure 5.3 Poorly installed MVHR technology can make it impossible for the inhabitants to use it properly

exercises which go through the motions of using the BPE methods available, gathering and analysing individual data, but without properly triangulating the results between physical and socio-cultural aspects, thus missing vital diagnostic insights. This can also happen when the BPE evaluator does not have enough interdisciplinary experience, or the BPE team is working in a multidisciplinary rather than an interdisciplinary way.

'Real world' action research

One way of overcoming the above 'blind spots', in relation to the performance gap and poor modelling, is to ground a housing BPE study using 'real world' research. This gathers quantitative data first and uses qualitative methods to help explain and interpret the data. Essential skills needed by a BPE evaluator to use this approach relate to the following:

- Writing a proposal
- Clarifying the purposes of the evaluation
- Identifying, organising and working with an evaluation team
- Choice of design and data collection techniques
- Interviewing
- Questionnaire construction and use
- Observation
- Management of complex information systems
- Data analysis
- Report writing, including making of recommendations
- Fostering utilisation of findings
- Sensitivity to political concerns

The evaluator needs to be in regular dialogue with the client, the inhabitants and other stakeholders in any study in order to learn from their insights. The evaluator can then intervene to help directly improve housing performance via the study. This is known as action research.[19] It is quite different from how most BPE is carried out, as it places the client and the inhabitants in the position of being 'co-researchers'. It can empower them to make decisions on how to undertake the BPE work, or to help with the design of the methods and data collection. Clients are generally willing to engage with this process once trust is established. Addressing inappropriate or unrealistic proposals put forward by the client, however, requires good negotiating skills.

A typical BPE cycle will:

1. Define the BPE study
2. Describe the situation and context
3. Collect evaluative data and analyse it
4. Review the data and look for contradictions

Most BPE studies simply make recommendations based on steps 1–4. Action research, however, involves some extra steps:

5. Tackle a contradiction by introducing a change
6. Monitor the change
7. Analyse evaluation about the change
8. Review the change and decide what to do next[20]

Action research is particularly effective during the commissioning and handover stage. There are excellent examples in housing BPE studies, where evaluators have helped housing inhabitants and managers to fine-tune the performance of heating and ventilation systems by spotting when control settings have been incorrectly set in the first place, or where technology is incorrectly installed. BPE evaluators can also build very close relationships with the inhabitants and work side by side with them to learn from them and help them learn about using their environmental controls more effectively.[21]

Commissioning the home

Commissioning requirements

Much of the reality of housing performance becomes apparent during the first six weeks after the handover of a home to its inhabitants. During this stage, they try to make sense of a bewildering array of new equipment and attempt to control it, often for the first time.

There are now standard requirements in UK domestic building regulations for installing, setting up heating and ventilation systems, and checking they perform correctly. This commissioning of services is vital if homes are to perform according to design intentions – and yet all too often it is found wanting, with systems poorly calibrated in relation to inhabitants' needs, or simply malfunctioning.[22]

Commissioning needs to encompass all service controls and systems, including fixed heating, lighting, water and ventilation appliances. It should include a check of all control settings against the manufacturer's recommendations, a visual inspection for any obvious errors in installation, faulty equipment or damage, and testing of equipment to ensure it is properly working and fully adjusted (e.g. carrying out airflow measurements to test mechanical heat recovery ventilation systems).

There is also a regulatory requirement in the UK to produce adequate guidance for the inhabitant, to be contained in an operation and maintenance manual, when installing these systems. However, these manuals are often written by manufacturing engineers who do not have the inhabitant's view of their home. This is because they never visit the specific context in which the equipment is installed, and the guidance they produce is far too generic as a result.

Handover and induction

Large housing developers often prepare a careful induction process, during which personnel meet with the new inhabitants, before, during and after handover, to explain how the home works. This hands-on approach is essential in order to contextualise the generic written guidance. However these developers can forget to ensure that the controls are properly demonstrated to the inhabitant, and smaller organisations may omit the induction process altogether. Moreover, the induction is only as good as the training of the demonstrator. Generally, customer care personnel do not have the skills to explain complex technology, and can inadvertently mislead inhabitants. In one large development, a home demonstrator explained to the inhabitants that their MVHR system would 'balance the heat in the house', when it clearly could not do this.[23] It is quite astonishing how little guidance or induction is legally required for the new inhabitants in ordinary private housing sales or lettings, given how much there is to learn. BPE needs to evaluate this area, which has a major impact on housing performance.

The key commissioning issues

Housing BPE is most effective when incorporated into the building process from the beginning, but it can still be useful when introduced at the commissioning stage in any newbuild or retrofit project. The evaluator can double-check any completed processes and feed back any issues to the client, suggesting interventions and giving clients time to address the issues before handing over to the inhabitants. If the BPE study is retrospective, it is still important to check the commissioning process if the initial BPE findings reveal unresolved issues.

The most critical commissioning to check in a housing BPE study relates to heating and ventilation system settings, and their performance, as there are often significant failures[24,25] here (see Figure 5.4). It is particularly important to include an evaluation of the inhabitants' understanding, as a perfectly commissioned system is no use at all if the inhabitants don't know how to use it. The poor commissioning of renewable energy systems[26] is an increasingly important area to tackle alongside digital control systems.

Figure 5.4 Airflow measurements are taken to check MVHR system performance

Inhabitant engagement with the home

Inhabitant impact

The impact of an inhabitant's engagement with their home provides another reality check for designers through BPE. An important example is the unanticipated 'rebound' effect[27] which occurs when inhabitants in a new low-energy home increase their previous level of comfort above the recommended room temperatures, simply because they know they will be saving money anyway and can afford to heat their home more. This can typically mean a 10–35% increase in energy use.[28] The results of the opposite 'pre-bound' effect can be equally dramatic. In one study of 3400 German homes it was found that inhabitants already consumed 30% less energy than anticipated, prior to moving into their new homes, meaning less energy saving overall than predicted.[29]

Bathing, showering, laundry and dishwasher use can dramatically increase water and energy use, with new needs emerging from new practices in response to new technology and changing societal norms.[30] In one study, the inhabitants of a supposedly 'zero carbon' home were carrying out the family laundry nine times a week.[31] Emerging technology is also reconfiguring the use of the home with the introduction of the ubiquitous digital media. Inhabitants are now demanding homes designed to accommodate the use of this in any room.[32] This increasing use of digital technology at home has seen an unpredicted rise in energy use that needs to be considered in BPE.[33]

A personal struggle with 'smart' equipment

Two key questions relate to effective engagement and control of the home. Firstly, do the installers have enough capability and incentive to explain and configure the system to match the inhabitants' needs through dialogue? Secondly, do the inhabitants have the understanding to be able to operate increasingly complicated systems effectively? Installers can often over-simplify selection, settings and explanation of heating systems to avoid having to deal with callbacks in relation to user misunderstandings. This 'scripting' of the system by installers can result in significant energy inefficiencies.[34]

My own fraught experience with the introduction of an innovative mobile thermostatic control unit in my own home in Sheffield illustrates this issue. The plumber said 'It's great. You can just take it from room to room and adjust your desired temperature for the whole home from anywhere.' What he did not point out was that there were already fully adjustable thermostatic controls on each radiator in the house. Here was the conundrum – if the temperature demand on the radiator valve in the living room was set low, but the mobile unit temperature was set high, this would signal a continuous demand for heat from the system even though the radiator valve was unable to supply the heat to the radiator. One wonders just how many households have wrestled with the same problem, and what effect this has had on energy use.

'Smart' homes

The advent of 'smart' homes, and technology services designed to 'think for' the inhabitant, has seen housing becoming ever more industrialised. This is particularly evident in the use of full-scale centralised mechanical ventilation systems[35] with ubiquitous air ducts and unfathomable heating or lighting systems. A 'smart' home can actually feel very disempowering for inhabitants, as the control is literally taken out of their hands.[36] If they are not able to use the controls properly at this stage, it can create a permanent problem. In one study, 32% of the inhabitants had simply switched off their ventilation systems altogether because they had received an inadequate explanation of their purpose.[37]

How much modelling and feedback to do

When to do modelling

Traditionally, housing designers do not engage with modelling beyond the minimum needed for regulatory compliance, with the home viewed as a relatively 'simple' building system, lacking the complexity of other building types. The design mistake has been to confuse spatial simplicity with performance simplicity. In many ways, a home is more complex to model than other building types, precisely because people live in it and have more control over it than the occupants in non-domestic buildings. If modelling is undertaken for compliance purposes only, then it is carried out at the detailed design stage. To use modelling for improving energy performance, introduce it early on in the outline design process to determine the best spatial form and site orientation for the home. For multi-storey housing, more sophisticated dynamic computer modelling is worthwhile, capturing thermal stratification between different story levels, for example.[38]

Linking modelling and BPE

Modelling either confirms intended housing performance, or refines it through iterative testing of different design options. It can investigate critical aspects of the design proposal, such as lighting, acoustics, ventilation or thermal design. Computer modelling is not very good at simulating the changing performance of all four of these aspects when a variable is introduced into any one of them – for example if outdoors is too noisy, a window will be closed, which will in turn affect both the ventilation and the thermal performance of the home. Here, utilisation of as-built, as-occupied and as-managed lessons from the post-occupancy evaluation of previous similar housing comes into its own early in the design process, providing a vital reality check for any modelling done.[39]

The housing design team need to ensure that whatever modelling approach is chosen, the BPE factors are also be built into the modelling from the start and that the BPE study methods tie in closely with the computer modelling approach. Otherwise, it is impossible to verify the simulation later on.

Levels of feedback

BPE feedback in housing projects can be used as '... a way of quality control in the more repetitive projects; as a necessary part of hypothesis testing in innovative ones; as a means of increasing awareness of chronic problems, changing requirements and emerging properties; and as a way of promoting fine-tuning and team learning'.[40] The important question is how much is enough? An affordable 'quick and dirty' study can often be enough to gauge whether a home is performing as intended. If problems surface in this initial study, however, a more bespoke approach will be required to try to pin down the underlying causes of the problems.[41]

Case study vs statistical analysis

How many homes make a BPE case study 'valid'? Even a single case study of one home can be worth a lot if it provides detailed insight into key problems for the housing sector, but it is often difficult to generalise results as 'typical' performance from any one particular study. It is more useful to have at least several examples of the same house type to compare in one case study, if possible, to rule out any one-off results relating to very particular circumstances.[42]

The use of meta-studies that combine data from multiple BPE case studies is even more fruitful in terms of pinning down generic problems and potential solutions. It is important to use exactly the same methods in each case study to ensure the validity of the meta-analysis. Far larger data samples related to housing BPE are needed, however, in order to produce genuinely epidemiological studies that cover thousands of households. Using this approach, definitive patterns and pathologies in housing performance can be determined in order to produce more robust models.[43]

An international epidemiological approach for understanding energy use in buildings is being developed through the *IEA Energy in Buildings and Communities Annex 70: Building Energy Epidemiology* report, with a publication date in 2020. This will complement the rich detail that a good BPE case study provides.

There is no need to develop the 'optimal' BPE study, which uses a huge variety of different methods and involves excessive amounts of data. No single study will ever fully capture the housing performance process, as it changes over time. It's far better to start with a 'satisfactory' BPE study design that is affordable, engage with the initial data, and only highlight the need for more work to be done if it is really needed. The primer at the end of this book explains how to develop this 'drill down' approach, starting with a 'light-touch' BPE study.

Optimising vs satisficing

Designers want to produce 'optimal' housing and use modelling and testing iteratively to try to achieve this. This is to meet inhabitants' needs with the most effective use of resources possible. But what does 'optimal' mean as a concept for housing performance? The question needs to be asked: whose 'optimal' is being chosen, and why? When it comes to defining 'optimum' comfort conditions, one person's 'warm' is another person's 'cold', depending on a wide variety of factors, and the same applies to all human perceptions.

Instead of designing homes to perform 'optimally', it is possible to offer a variety of different ways to meet any particular need more resiliently, so that if one way fails, another is available. This introduces a degree of useful 'redundancy' in housing.[44] Recent research shows that it is better to design housing to offer a broader range of 'optimal' temperatures[45] with more opportunity for localised personal body heating elements and control options[46] (see Figure 5.5).

Inhabitants usually make decisions to achieve a 'satisfactory' answer to their needs, with minimal effort, time and resource, rather than aiming for the best outcome, which may entail greater effort. This 'satisficing' means that they will not necessarily engage with a home in the way that the 'optimising' designer intended. They may well tolerate heating, lighting or acoustic conditions that are 'good enough' and go no further. Interestingly, house builders also engage in satisficing, as they have to balance price against performance – houses are thus very rarely truly 'optimised'.

Housing BPE studies need to take the concepts of satisficing on board within their research design, and adjust the performance parameters accordingly. This involves finding out from the inhabitants what is satisfactory for them, rather than imposing assumed 'optimal' conditions on them. In some cases, however, inhabitant satisficing may have to be overruled. Good ventilation is critical for good health, and inhabitants' senses are not always the best judge of this. Equally, when people get older, their sensory powers can diminish, and their satisficing may well be based on poor sensory feedback.

The approach to modelling and reality described in this chapter highlights how a BPE study can inform housing performance models and design through feedback. There is, however, a good case for carrying out repeated BPE studies of the same project over time to improve housing continually, as discussed next.

Figure 5.5 It is useful to evaluate whether a home is 'optimal' or flexible – people need ventilation and heating choices

SIX
LONGITUDINAL FEEDBACK AND DESIGN ITERATION

" ... the home needs to be understood as an ongoingly changing configuration of people, things and processes. "

Pink, et al., 2017,
Making Homes:
Ethnography and Design, p 15

Living in a home is a continuously changing process. Inhabitants' needs evolve over time, and most homes have several sets of successive inhabitants over their life cycle. As inhabitants' circumstances change, the physical building also needs to be able to respond to this. Additionally, different elements of the fabric wear out and need renewal at different time intervals. Designers and housing developers who revisit their projects regularly over a long period will build up a deep understanding of how well their design products are really performing and adapting over time. Too often, the housing BPE box is ticked, and there is relatively little learned beyond the initial study and immediate subsequent fixes applied to the product design. This chapter completes the theoretical section of the book, setting out the role of longitudinal studies in improving design iteration, and the need to consider BPE through the complete life cycle of a building. This relates to the enhanced internal organisational and personal learning and capacity-building that can develop once BPE becomes a structural part of organisational processes.

The following aspects are covered in this chapter:

- Housing life cycles and feedback
- Evaluating housing through time
- Organisational learning over time
- Reflective practice for designers

Housing life cycles and feedback

The life of a home

The simplest definition of a housing life cycle is from the moment a home is completed until it is demolished. In reality, however, this very much depends on who decides what the life cycle boundaries are. A mortgage lender will typically define the durability of the newbuild home as 60 years in the UK. Homes, though, can often last for hundreds of years given the right care and attention, and there are many different component life cycles operating within any one housing life cycle.

Fast and slow change systems

There are fast and slow change systems within the overall life of a home. Stewart Brand offers a helpful diagram in his book *How Buildings Learn*,[1] which explains the time and spatial relationship between six key housing life cycles – stuff, space, services, skin, structure and site (see Figure 6.1).

Construction and service products have limited warranty times, even though the degree of care can make a huge difference in how long these products actually last. This is important, because housing BPE benchmarks performance against the assumed life expectancy of different housing elements and their service requirements. Check these assumptions against reality before commencing any BPE study.

The six physical life cycles illustrated below also depend on the social changes that go on within a home over its lifetime. As inhabitants grow older, they may expand the household with children, or by moving elderly parents in with them. They may also have changing needs as they get older, or suffer from a disability. Pink et al.[2] discuss this trajectory of a changing home and how to evaluate the future-enabled home in terms of the 'changing configuration of people, things and processes'. People carry out home renovations for all sorts of personal reasons which are unanticipated by the designer, often prioritising aesthetics over energy efficiency measures[3] – a factor frequently overlooked by BPE studies.

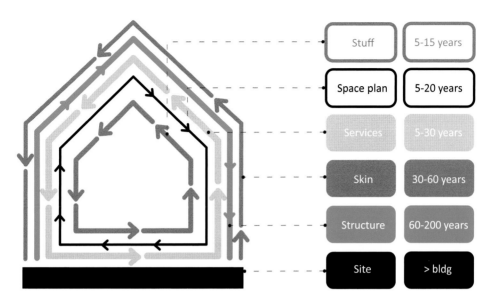

Figure 6.1 Fast and slow change systems in a building

Using longitudinal housing BPE to improve design

BPE does not predict the future, but it can certainly provide insight into the effectiveness of the adaptability designed into any home, which may help future designers to better understand this phenomenon. Understanding how well a home has accommodated evolving lifestyles (its 'interactive adaptivity') has yet to be captured in housing BPE methodology, which tends to be based on a static view of housing needs. This is where long-term housing BPE can really pay off. By comparing an initial housing BPE study with another one carried out a few years later, it is possible to analyse how well a home meets changing needs, and inform a more effective design process.

This type of 'double' study can also help with understanding the emergence of new technologies and their performance in the home. However, these technological objects also have their own digital codes which '… alter the material, social and spatial relation of the home in new ways; they offer members of households new affordances to undertake domestic living differently'.[4] Guidelines for encouraging good emergent properties in housing are helpful when considering how to evaluate these more changeable aspects. Simple low-tech features are generally more resilient and tend to cope better with emergence than active, complicated ones.

Longitudinal BPE studies can also help to highlight how initial capital expenditure (capex) directly affects operational expenditure (opex) and maintenance costs. BPE is one of the few ways to gather evidence on this to inform future decisions in terms of initial cost outlays for housing developments and housing life cycle costing more generally. One example is short-sighted low-cost specifications for windows that result in significant maintenance and repair costs later on, in addition to the environmental cost of the waste.

Timescales for BPE feedback

A longitudinal housing BPE study is highly appropriate as part of any planned major maintenance event (e.g. when any external fabric or heating systems are due to be replaced). A study done after the event will show how well the upgrade/replacement activity has worked. Undertaking a longitudinal study as part of the selling/rental transaction for a home can reveal how different inhabitants engage with the home and how flexible it actually is. The study can become part of the transaction as evidence of performance in reality – providing much better information than the routine predictive RdSAP and EPC reports carried out in the UK. It is important to be clear about the precise purpose of the longitudinal BPE survey, to determine the best time to carry it out.

Evaluating housing through time

Revisiting Kincardine O'Neil: a personal case study

In 2002, as a researcher attached to the University of Dundee, I undertook a six-month housing BPE study of 14 innovative low-energy houses in the remote village of Kincardine O'Neil, Aberdeenshire, for Communities Scotland. The compact, single-storey, all-timber housing development was designed by Gokay Deveci Architects and completed in 1999 (see Figure 6.2). These houses were the outcome of a research project that sought to address the need for genuinely flexible and affordable rural housing in Scotland. I was then given the rare opportunity to carry out a second follow-up six-month study in February 2004 to note any further changes that may have taken place.[5]

Figure 6.2 Evaluating innovative housing over five years – Kincardine O'Neil

The second study aimed to establish:

- the ongoing level of heating bills and energy costs compared to normative measures
- the ongoing effectiveness of the flexible design concept
- the overall tenant satisfaction with the houses and level of rent
- the level of turnover, achievement of targets in allocation policy and any management issues
- a comparison of the ease, speed, and cost of construction with a standard housing development
- the extent to which the development contributed to the economic sustainability of the village.

A full documentary analysis of drawings, specifications, energy calculations and maintenance records was undertaken prior to semi-structured interviews and informal home tours with 12 out of the 14 households for the first evaluation. Five households were re-interviewed for the follow-up evaluation. The client, architect and contractor were also re-interviewed to ascertain their views on the housing development over time. Fuel bills for a one-year period were compared to all the other data. Representatives from local businesses were interviewed to understand the wider impact of the housing development on the village.

A key finding was the failure by the property owner to deal with severe thermal and acoustic issues highlighted in the first study. However, he planned to replace the undersized mechanical ventilation units, as they contributed to mould growth in some shower areas, indicating what his real improvement priorities were. The design quality of these homes generated a long-term 'forgiveness' factor in relation to these particular challenges, with an even higher level of tenant satisfaction in the second study (100% compared to 82.5% previously). The external aesthetics were much more acceptable to the tenants interviewed in the second study, with several commenting on the softening of the timber roof shingles and matured soft landscaping. However, only one household had taken advantage of the deliberately flexible design of the home, having installed a staircase at their own expense to make use of the attic space. Others did not do so on cost grounds. The long-term study helped to reveal the relative ineffectiveness of the planned adaptation measures.

The impact of the development on the economic and social life of the village proved to be minor, as most tenants tended to go elsewhere to meet their needs. Just under half of the households moved out between the two studies, indicating a very high turnover and highlighting the need for careful handover procedures and guidance. These were still not in place at the end of the second study. Just how important these particular elements are became apparent in a much larger longitudinal study.

Longitudinal housing BPE as a meta study

Together with a planner, Nick Williams, I wrote the first guidebook on *Sustainable Housing Design for Scotland*.[6] This contained 13 innovative housing case studies demonstrating the use of low-energy technologies, including super-insulation levels, passive ventilation systems (Passivent), breathing walls, sunspaces, communal heating, solar thermal panels, ground-source heat pumps and MVHR systems. Basic post-occupancy evaluation (POE) was undertaken on each development, including documentation analysis, home tours, interviews with design and housing management teams, comparison of energy costs with predicted costs where possible, and discussions with tenants. Seven years later, we revisited the sites and re-interviewed the key actors to establish what had changed over time, specifically in relation to the use and understanding of the new technologies installed (see Table 6.1).[7]

Table 6. 1 Housing technologies evaluated over time

Low-energy features 1999 revisited	Would provider use feature again?	% providers who would use again
High insulation	Yes, if affordable	100
Passive ventilation	Yes	100
Breathing wall	Yes	100
Sunspaces	Yes, amenity only	66
Communal heating	Yes	66
Solar panels (water)	Yes, if affordable	100
Ground-source heat pumps	Yes	100
MVHR	No	0

In one case, district heating pipes in a 23-storey refurbished tower block had corroded due to the rust inhibitor leaking out of valves. Over £800,000 of repairs were undertaken. The whole pipework had to be replaced because of the difficulty of locating valves buried within floor screeds and behind panelling. Vital engineering drawings had been 'lost', making the repair process much more difficult. Long-term issues in other cases highlighted the difficulty of dealing with emergency failures in ground-source heat pump and MVHR systems when the manufacturer was remote, and both servicing and parts were hard to obtain quickly. Badly eroded tenant and housing provider confidence meant that the MVHR system was removed, with no plans for using MVHR again.

The second study revealed that many tenants were heating uninsulated sunspaces in the winter by leaving the connecting doors to the living room open – a classic mistake (see Figure 6.3). Tenants admitted they had little understanding of how sunspaces worked, and housing managers confirmed that successive tenants had not received any written guidance. There were no remote sensors to alert management staff to system breakdowns. Instead, they relied on vague reporting of 'heating failure' by tenants that proved difficult to diagnose.

All these issues only became known through the longitudinal study, revealing major consequences from what appeared to be a minor concern – guidance and continuity. It also became clear that two groups had very different views of the installation and use of low-energy features. The housing development officers focused on the need for good communication between the different sections of their organisations to ensure that feedback on the performance of these features was built into the procurement process. By contrast, the housing managers focused on the needs of their tenants and communicating mainly with them. This raised further questions. What feedback was really going on in these housing organisations? What were they learning, and how?

Figure 6.3 Tenants heating uninsulated sunspaces is a classic BPE finding

Organisational learning over time

Longitudinal BPE enables inhabitants, housing developers and design teams to benchmark the performance of their homes and technology over time. This allows them and policymakers to reflect on any changes that have taken place in their homes, to learn, and to build on the capacity to deal with changes by highlighting long-term patterns and trends related to information management, technological innovation and inhabitant engagement. Finally, longitudinal BPE helps to articulate and store the tacit knowledge that so often gets lost when personnel move on, and thus makes organisations more resilient in the face of change.

The BPE learning process

Learning can be defined as 'the process whereby knowledge is created through the transformation of experience'[8] through a personal reflection on direct experience, forming ideas (or concepts) based on this, and testing these in practice, which in turn creates new experiences on which to reflect.

Organisational learning in practice means individuals acquiring new knowledge, passing this on to a group level for others to interpret and understand, and developing an integration process that enables everyone to agree on a common understanding. This enables the development of new formal routines and procedures. The feedback cycles as shown below.[9] provide a useful means for understanding how BPE can continuously inform the broader housing development life cycle as part of a long-term process (see Figure 6.4).

BPE lessons identified during the briefing, design, implementation and initial inhabitation stages can be immediately fed forward into the housing project (level 1). Where the lessons cannot be applied directly, they can be retained at an organisational level for deployment in future developments (level 2). Following the housing performance evaluation, the lessons from this can be applied immediately to the next identified development, or again stored within the organisation for future use. The housing developer and design team gradually build up a knowledge bank to produce an evidence base for decision-making (level 3). The key to embedding organisational learning, however, lies with levels 4 and 5, through knowledge management and consolidation in practice of the lessons learned. How one housing developer responded strategically to longitudinal BPE feedback is considered next.

Crest Nicholson: organisational learning over time

I first encountered Crest Nicholson plc, a major UK housing developer, via their Group Sustainability Executive, Elizabeth Ness, in 2008. She was very serious about Crest Nicholson developing as a learning organisation and using housing BPE in relation to levels 1, 2 and 3 of the organisational feedback cycles illustrated below. We completed a BPE study of 200 homes at their Avante development in Kent in 2011[10] (see Figure 6.5). The most interesting part of working with Elizabeth and her sustainability manager, Julia Green, was discovering where their organisational processes needed strengthening at levels 4 and 5 as a result of the insights generated by the BPE work on Avante. These two entrepreneurial 'sustainability champions' created a very different dynamic in terms of the pan-organisational response

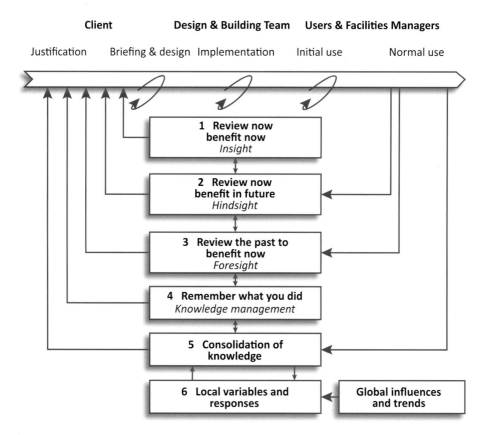

Figure 6.4 Organisational feedback cycles for the BPE learning process

compared to previous housing developers I had worked for. They were able to promote feedback from the Avante BPE project across the organisation at the highest level.

The final Avante BPE project review meeting led to the development of an organisational action plan for future improvements that strategically influenced the next generation of standard house types across the organisation. This form of review provides a textbook example of how to use housing BPE effectively for organisational learning. The developer introduced 'business improvement workgroups' to develop new working practices, and committed to a new organisational strategy for ironing out defects. This included creating a new panel model for engaging design consultants on a long-term basis, embedding learning from BPE, as well as providing BPE feedback to key suppliers.

The two sustainability champions also began to explore the Soft Landings framework in terms of how feedback could be deployed within the organisation itself. The developer strategically engaged with all five stages of Bordass's feedback cycles (see Figure 6.4), with the sustainability champions working closely with key managers across the organisation. Ultimately, this enabled Crest Nicholson to interrogate its business practices, understand the root organisational causes of various problems and attempt to solve these via institutional changes. Ten years later they were still improving their organisational processes, but this time using a form of ethnographic self-critique of their quality management routines via an embedded architect/PhD student. This is a classic example of continuous organisational learning over time, which uses housing BPE not only to learn how to design better homes, but also to successfully question the underpinning of their learning processes.

Figure 6.5 Crest Nicholson's Avante housing development

Reflective practice for designers

Reflective practice in relation to BPE

Organisational learning using housing BPE can also help design practices to improve their processes and design products using a similar approach. However, designers face conflicting values, uncertainty, unique situations and unpredictable change when engaging with a new project, and cannot always resolve these using existing knowledge and rule-bound procedures. In an important book, *The Reflective Practitioner*,[11] Schön describes the notion of 'reflection-in-action'. This moves designers beyond existing known theories and procedures, and allows them to access their vast tacit knowledge to solve problems holistically. An artful skill is deployed in the act of actually doing something, which could be as 'simple' as doing a design diagram at a community consultation meeting, for example, or sketching on a notepad to understand why a hidden detail is not functioning as it should. The danger is that the design outcome remains at this level – tacit and untested. Formal housing BPE can capture this tacit knowledge-in-action as an act of explicit testing and learning which others can draw on.

Remaining open-minded, demonstrating a willingness to probe for, discover and learn from mistakes, to be able to think 'out of the box' – all are key personal attributes which lead to a successful reflective practitioner being able to learn from and embed BPE as a habit and ethos. The willingness to objectively assess and evaluate how well a housing design project has succeeded in reality takes a lot of moral courage and not a small degree of professional humility. This is difficult in the face of architectural education that still encourages individualistic 'starchitecture' rather than teamwork, and has no requirement for BPE training as part of the professional curriculum (see Chapter 9).

Developing housing BPE in a design practice

A design practice needs a belief in learning through continuous design evaluation, at *all* stages of the housing life cycle. Without this belief, housing BPE cannot be fully embedded within a practice as an evolutionary learning process. Beyond this, a practice needs to be prepared to continuously prioritise dedicated time and budget in order to build in BPE policies and procedures within the organisation on a cyclical basis. This means using methods and techniques to help individuals and groups to reflect on, and share, their BPE experiences and actions in order to develop a process of continual 'single-loop' (refining existing practice) and 'double-loop' (developing new concepts) learning at every organisational level. Design principles, strategies and specifications can then be continuously re-informed from real-life feedback related to building performance, rather than remaining speculative. This ethos of reflective practice requires the development of robust practice-level knowledge-exchange systems using critically reflective workshops, best-practice notes and design procedures that directly incorporate learning from BPE into new drawings. Systematically embedding this evaluative approach into the DNA of a design practice can have spectacularly positive results, as demonstrated by Architype Architects. BPE has continuously refined and improved the performance of their projects and reduced future risk and liability for this design practice.

Embedding housing BPE as reflective practice

Another practice that has successfully embedded BPE into their housing work over the years is Richard Partington Architects. Ironically, Richard's first exposure to intensive BPE followed on from narrowly missing out on the Elm Trees Mews housing competition for the Joseph Rowntree Trust:

'We were really fortunate not to have won that. If we had, it [would have been] catastrophic technically. By not winning it, but doing well, I was asked to join a research group. That led to working with Kevin Lomas, and particularly Bob (Lowe) and Malcolm Bell at Leeds. Without planning it, we got access to this knowledge sharing, practice/academic/research overlap thing. It wasn't structured in any way, but we just benefitted hugely from exposure to those sort of people in that environment.'

Later on, the practice won a much larger competition to develop a new suburban housing community in Derwenthorpe near York, which led to further BPE development:

'On Derwenthorpe, the scheme that we'd won with, Rowntree said "Well. Given the Elm Tree Mews experience, why don't we build some prototype houses and test them?" That's when the BPE really kicked in and that's when the relationship with Malcolm Bell and Jez Wingfield was, in design terms, on a weekly basis. They then said "Why don't you get someone in-house to do all the thermal bridging calculations?" Malcolm's first observations on Derwenthorpe were that geometrically it was far too complex, and had we considered all the heat loss through these geometrical junctions. We had no idea.'

The practice went on to continually reflect on their design work, which produced an unexpected outcome:

'... it was never intended that Derwenthorpe would be delivered by one practice, but by promoting this culture of thinking, learning, re-appraising what we have done, and being honest about things that haven't worked, celebrating the things that have – we've [been] working on it now for 10 years.'

Developing BPE as a practice culture can encourage repeat work, which pays dividends all round.

The challenge of long-term reflective practice

Designers may believe they are already engaged in 'reflective practice' when evaluating their housing projects. They may revisit them, talk to the inhabitants, note defects and think about what to do next time. This is what the successful Dutch architect Marlies Rohmer did when she informally revisited every one of her projects after 25 years in practice, chatting to the occupants and associated building experts.[12] She notes that architects need to be more critical of themselves and be brave enough to treat a revisit as a form of therapy in order to be able to turn failures into lessons.

Probing more deeply, however, reveals that sometimes a more critical review of her own work is lacking, possibly because no formal BPE process was employed. When challenged by residents about the difficulty of reaching the exterior cladding on their floating homes to clean algae off, for example, Romher believes that it should have been explained to them before they bought the homes that a water dwelling needs more maintenance. There appears to be no learning from the revisit about the need to design floating housing that inhabitants can maintain more easily. In other visits, however, Rohmer clearly does take on board lessons from feedback. Many designers will relate to these housing 'revisits' and the wistful view of failed design due to 'failure to educate the occupants'. Educating inhabitants, however, is no substitute for undertaking BPE and then being willing to let go of a bad idea, based on evidential evaluation, and improving the design. The next section of this book explains the training needed in order to do this.

TRAINING
FOR
FEEDBACK

SEVEN
FEEDBACK TECHNIQUES

"To achieve the greatest uptake, the ideal feedback technique would be simple to use; widely applicable; robust but comprehensive; and cheap, and quick and easy to operate."

B. Bordass & A. Leaman,
'Making feedback and post-occupancy evaluation routine 1: A portfolio of feedback techniques', Building Research & Information, 33, 2005, pp 347–52.

If doing housing BPE is so complex, what is a realistic way forward to capture the necessary information and insights that is both affordable and effective? How does one choose from the vast array of available techniques? This section of the book sets out what and who is needed to carry out successful housing BPE studies. This chapter describes an approach for using current housing feedback methods believed to be most useful, with more information in the primer at the end of the book. Chapter 8 discusses emerging techniques that can augment these studies. Chapter 9 describes current educational and training approaches for BPE.

Homes are very intimate and private places compared to non-domestic environments. An extra set of skills is required to gain the trust of inhabitants and access to their feedback to help improve homes for a wide demographic. An appropriate combination of BPE methods and techniques is critical to understand housing performance in terms of the interplay between the inhabitants' practices, the wider socio-cultural context, and the physical behaviour of the home itself. Housing BPE evaluators, whether they are in-house or external experts, need to keep abreast of the latest developments in this field, as well as understand the full context of a project, to be able to select the appropriate methods and techniques.

The following aspects are covered in this chapter:
- Standards and benchmarking
- Overview of techniques: what to use, when?
- Auditing
- Physical evaluation of housing fabric
- Monitoring of environmental qualities
- Surveys, interviews and focus groups
- Ethnography

Standards and benchmarking

Existing standards and BPE
The requirements of legislation alongside the needs of inhabitants define key housing performance criteria and standards for BPE. The criteria cover environmental, health and safety, security, comfort, functionality, efficiency, social, psychological and cultural aspects. There are still numerous gaps in housing performance legislation related to indoor air quality, lighting quality, water consumption, usability, ventilation and heating system feedback for the inhabitant. Legislation governing areas such as air-change rates and overheating levels is not enforced due to the lack of training of building control officers, and their depleted presence in the UK. BPE addresses the vacuum created by this lack of enforcement of standards. It also highlights the need for more bespoke solutions to suit specific circumstances.

At the same time, there is a bewildering range of existing standards with BPE components against which to benchmark the performance of housing, typically produced by the International Organization for Standardization (ISO), the European Committee for Standardization (CEN) and the British Standards Institution (BSI). The American Society of Heating, Refrigerating and Air-Conditioning Engineers (ASHRAE) and the UK Chartered Institute of Building Engineers (CIBSE) also provide standards and guidance. Key UK standards are set out in the Primer in relation to relevant BPE methods described below.

Benchmarking housing performance in BPE

Organisations using benchmarking compare their results with best practice by others, learn from this, and stretch performance targets over time. Benchmarking is a well-tested method for improving performance, particularly in relation to quality, time and cost.[1] Comprehensive benchmarking datasets for housing BPE studies are, however, still surprisingly hard to come by.

The following international standards give credits for housing post-occupancy evaluation (POE) that can be benchmarked:

- LEED for small housing developments[2]
- LEED for larger housing developments[3]
- BREEAM[4]
- Living Building Challenge[5]
- WELL – health and wellbeing only[6]

These standards are discussed more in Chapter 10.

Various government agencies and NGOs around the world collate housing performance studies, and research institutions compile their own datasets related to various initiatives from time to time. However, it is unlikely that a coherent set of data will ever develop without legislation, given that house builders in the UK generally respond more to legislative rather than voluntary drivers,[7] and UK architects also feel that legislation is required to enable BPE to become routine.[8]

Emerging standards

The BRE Home Quality Mark is a relatively new housing performance standard in the UK.[9] All the POE credits are optional, and there is no mandatory requirement to undertake this activity. Just three POE methods are specified for a 'light-touch' study – analysis of inhabitant feedback, energy and water bills – all compared to design intentions. There is no requirement for a home tour, meaning the exercise can be done remotely, with little understanding of the complexity of the home context in reality. This needs improving.

What happens if no standards exist for evolving circumstances, such as climate change? In this case, using a group of experts, inhabitants and managers to generate new performance criteria can help to futureproof BPE studies.[10] A key example here would be to build in future climate prediction temperatures, such as those offered by CIBSE, for benchmarking BPE housing performance to identify future summer overheating of homes and associated health risks.

Overview of techniques: what to use, when?

Using a diagnostic approach

Although 10% of architectural practices in the UK offer POE as a service, only about 3% actually deliver it regularly on housing projects.[11] The latter get substantial benefits in terms of developing as a business.[12]

The best approach to housing BPE is to start with an affordable and manageable process, generally known as a 'light-touch' BPE study, based on four questions:

1. Is the home physically performing as expected?
2. Are the inhabitants' needs met in their home?
3. Are there any physical and/or social problems that need solving just now?
4. How can we improve our housing for the future?

The 'drill down' approach starts with this investigative light-touch BPE study, which is within the grasp of most qualified architects, and only goes more deeply if there are problems raised which require further expert diagnosis. Forensic investigation is required where problems are particularly complex. If the light-touch study reveals that a home is performing as expected, inhabitants' needs are met, and there are no particular problems, there is really no need to do more BPE just for the sake of it. It is far better to save the money for where problems are likely to emerge, for example for an innovative housing project.

Each level of BPE has a range of methods and techniques associated with it, which are increasingly more intrusive. There is no generic approach, as different projects and contexts will probably need slightly different combinations of these methods. Equally, the light-touch approach should not become the industry norm, to the exclusion of deeper investigation where needed.

The primer at the end of this book highlights all the relevant source references related to each method and technique.

Light-touch level

Basic feedback from a housing project can yield a significant amount of insight, when carried out by the design team. This could be a single site visit, with a guided tour through a sample home and a conversation afterwards with key actors related to the project (typically, managers and inhabitants).

When followed by a short informal group talk within the practice office to work through the issues raised, this can stimulate useful information for future design development by providing evidence for assumptions about what works and why. If this is done on a routine basis, deep learning can accrue within a practice over time.[13]

A slightly more formal light-touch BPE study includes the following:

- A brief review of key design, construction and specification documents related to the project intentions
- A basic energy and water use survey
- A simple survey of the inhabitants
- A short home tour with inhabitant/developer/design team discussion in at least one of the homes [14]

A spot-check of environmental conditions (temperature, humidity, light and sound levels) can be made using a simple multi-meter to correlate with any initial observations made during the home tour (see Figure 7.1). As thermal imaging becomes increasingly affordable, it is worth making a brief camera survey of at least one home to identify any obvious heat-loss issues (see Figure 7.2). Ideally, energy-use data should be downloadable from smart meters and sub-meters already designed into the housing project. Where these are not available, it may be necessary to add in additional energy sub-meters, but only if energy use appears to be excessive from the basic meter readings.

Figure 7.1 A spot-check using a 4-in-1 environmental multi-meter can reveal a variety of sensory issues

Figure 7.2 Thermal imaging is increasingly affordable, and a brief survey of one home is worthwhile to locate heat loss and air leakage

At this initial level, the skill of the individual BPE evaluator, whether they are based within the practice or externally recruited, is paramount. With a high degree of knowledge, skill and experience, it is possible to rapidly comprehend and address many of the structural, construction or services issues that can arise during the home tour. The evaluator also needs to be alert to issues relating to good health and wellbeing, clean water and sanitation, affordable and clean energy, sustainable consumption and production, and climate action and biodiversity, as key housing needs identified in Chapter 4.

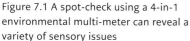

Diagnostic level

If problem areas are indicated with causes that are not easily understood, it will be necessary to move on to more intrusive methods for diagnosing and unravelling what may be a 'wicked problem' – one with multiple factors to address and that requires expert help.

The airtightness of the home may need to be re-tested by a suitably qualified technician, along with a re-testing of renewable energy, heating and ventilation commissioning processes (see Figure 7.3). The SAP calculation should be double-checked, in case the prediction was wrong in the first place. Detailed analysis using energy sub-metering may be necessary. Partial monitoring of temperature, humidity, lighting, acoustics, CO_2 levels and other toxins may be useful in relation to certain areas of the home – most likely the living room, bedroom, bathroom and kitchen. It can be useful to carry out short-term heat loss analysis through key elements, with more comprehensive and detailed thermal imaging related to a full photographic survey of the construction externally and internally, as well as comparison with the relevant construction details – as designed and as built. Suitably trained and qualified personnel will need to carry all these tasks out.

Figure 7.3 Re-commissioning MVHR systems can reveal significant failures, either in the original commissioning or the system itself

Always combine these physical activities with further expert-led interviews and walkthroughs to try to understand in more detail how inhabitants are engaging with their homes and identify any physical problems emerging. As the causes of housing performance issues are often upstream of the inhabitation stage, it is also be useful to evaluate the home handover and induction process at this stage – which an experienced architect or engineer can do. Compare the findings from all the methods used against the design intentions as manifest in drawings, details and specifications.

Forensic level

Further issues and underlying causes may require a forensic approach to understanding the performance of the home, with physical and social science BPE experts working with the design team. This includes an extended monitoring period combined with ethnographic work involving inhabitant diaries or logging and videoing events, and defining occupancy patterns. This can triangulate the various complex factors at play in a home that feed into deeper problems.

An expert-led thermal comfort survey can provide a more accurate understanding of inhabitants' perceptions of temperature, humidity and wellbeing. A usability survey can reveal any issues related to the ease of use of service equipment in the home, and consequential impacts on health as well as sustainable consumption of water and energy. Focus-group exercises led by a social scientist can also help to draw out agreed underlying issues in more detail. A co-heating test organised by experts can conclusively establish the overall heat loss through the fabric of the home. The use of probes to identify moisture movement through the fabric can help to understand how moisture is moving around the home, particularly in relation to dampness and subsequent mould growth which could lead to health issues, as well as degradation of the housing fabric. Finally, it can be useful to identify what doors and windows are being opened and when, as well as who is where and when, using sensors to capture movement over a given period of time.

The order of BPE activities is important – always audit before carrying out any physical monitoring and then establish some physical measurements before approaching the inhabitants for their views. This way you can cross-check what was supposed to happen (the audit) with what actually happened (physical and social surveys). Identifying which methods are relevant to each of these three levels of investigation requires an understanding of their exact purpose as well as the resource requirements for each one, as discussed next.

Auditing

Documentation review

The documentation related to a housing project embodies and preserves the thinking and actions of the project over time. Understanding the documentation, how it has been stored, and whether any verification of the documentation has been undertaken, are key elements of a BPE audit. This provides a base line to compare with other findings. An audit also maps out how well documented the housing project is in the first place. If the documentation is poor, it is likely that there will be performance issues. A basic audit can be carried out by any design team professional, but they should ideally be independent from the project design team, for a 'fresh pair of eyes'. Key documents to audit are set out below.

Health and safety strategy and standards

These are set out in the health and safety file for the housing organisation and project, if there is one. The UK CDM regulatory requirements should be checked for compliance if any issues are raised in relation to this area.

Spatial quality and layout strategy

Space standards should include storage and adaptability, volumetric quality, and external space and environment quality, as well as key location, site and layout plans and section drawings, and any underlying assumptions/calculations related to spatial quality. Note any alterations compared to design intentions, and check against any relevant issues arising.

Construction strategy

This includes construction standards, outline specification of construction and materials, and embodied energy calculations, as well as key construction plans, sections and detail drawings, and any underlying assumptions/calculations related to these. Check these against any issues arising and against final drawings/specifications, noting any changes and the reasons for this.

Environmental systems and 'smart' control/metering strategy

These include water, energy, ventilation, thermal, lighting, acoustics and biodiversity standards, as well as service specifications: plans, sections, detail drawings and any underlying assumptions/calculations related to these. These should be checked against any issues arising, and against final drawings/specifications, noting any changes and the reasons for this (see Figure 7.4).

Maintenance strategy

This includes a whole-life maintenance plan, CDM standards related to ease of access, and relevant maintenance logging of events. Compliance should be checked against any issues arising.

Sustainability strategy

This includes sustainability standards, including climate change objectives, cyclical resource use, wellbeing, regenerative development, biodiversity, greenhouse gas emissions, etc. Compliance should be checked against any issues arising.

Handover and induction strategy

This includes guidance documentation and handbooks for inhabitants in relation to how to operate their home, guidance for housing personnel carrying out home demonstrations, and induction for inhabitants. Check these against any issues arising.

All documented strategies should include climate change adaptation requirements and should be checked for this.

Equipment and resource use audit

Water survey

Monitoring water meters can reveal if water use is excessive. Where meters are not available, water bills related to volume used can help to evaluate the problem. It may be

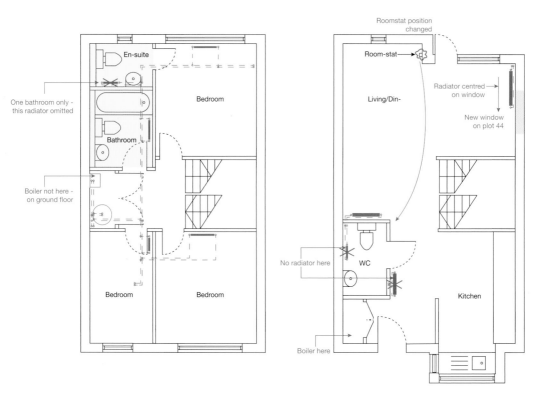

Figure 7.4 Specifications and plans are often deviated from during the construction process

useful to install a water flow meter for expert forensic investigation to cross-relate water use with energy use. Carrying out an audit on any equipment using water can also cross-relate water use with the claimed level of water-saving efficiency of the equipment.

Usability survey

This questionnaire-based tool, developed by the author working with other researchers, involves a half-day survey of all environmental controls in a home and how easy inhabitants find them to use, It should be checked against an audit of the specifications and drawings to highlight any issues.[15] This activity can be carried out by any member of the design team.

Physical evaluation of housing fabric

The physical performance of a home sets out the reality benchmark against which to judge the project intentions and inhabitant engagement. No housing BPE study is complete without one.

Physical evaluation methods

The following evaluation methods involve increasing levels of complexity and intrusion in the home. They can be used individually or in combination, depending on the issues emerging from the initial light-touch study, and the particular requirements of the BPE at any stage. The relevant person to carry out the work is shown in brackets at the end of each item.

Construction audit

Observation of construction and housing fabric via walkthrough, noting anything unusual compared to the construction strategy audit (any design team member). In extreme cases, where there is building failure, it may be necessary to undertake structural testing of elements in a laboratory.

Photographic construction survey

This type of external and internal survey is not particularly disruptive and can be undertaken while the inhabitant is at home. However, some furniture or other elements such as leisure equipment or curtains may need to be moved in order to enable a full survey to take place internally (any design team member).

Thermal imaging survey

Thermal imaging is similar to undertaking a photographic survey and is not particularly disruptive. Do both the surveys at the same time to enable accurate comparison. This process usually takes no more than a few hours for an average home. Again, some items may need to be moved to allow a full survey to take place (trained design team member or expert).

Air permeability

Re-testing the airtightness of an average home can be done in half a day, but it is relatively disruptive to inhabitants, as the home must be sealed for this purpose, with no one entering or leaving the home during this period (qualified technician).

Fabric heat-loss testing

This testing is quite disruptive to inhabitants as it may involve the placement of heat flux measurement pads on various walls, floors and ceilings for a minimum of two weeks in order to measure actual U-values (expert).

Co-heating test

This test involves a lot of disruption to inhabitants and is best carried out on an empty house (see Figure 7.5). It usually needs a minimum of one week's effective monitoring data obtained over a minimum three-week period. Some joints may need repairing if there is any slight movement in the fabric during the test. A more recent method known as 'QUB' claims to produce reasonable results for establishing the total heat-loss co-efficient for a home in just one night (expert).[16]

Moisture movement through the fabric

This process can be quite disruptive for inhabitants, as it may involve installing moisture probes deep into the fabric of a home, as well as equipment linked to remote metering and sampling. Some surface repairs may be required at the end of the process. Consider the home as a whole when diagnosing moisture movement, not just single elements (expert).

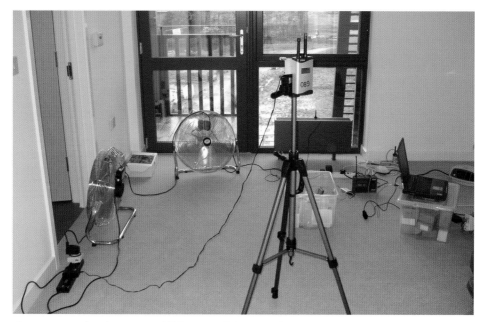

Figure 7.5 A co-heating test establishes overall heat loss from a home, but it is disruptive

Re-commissioning of services

Re-commissioning of home service systems in a BPE study requires validation of their proper installation, observation of start-up and component testing for heating and ventilation systems, and short-term diagnostic monitoring of dynamic interaction between these systems and inhabitants, in order to identify and resolve any problems. Identify no-cost, low-cost and capital-intensive improvement measures as part of this exercise. The following method should be applied.

Document analysis

Gather all relevant standards, manuals, drawings and specifications, and check service equipment against these. This includes any available database used to keep track of the original commissioning activities and information, the commissioning plan, checklists for each piece of equipment, test plans for the same, testing of equipment, any original deficiency/problem reports, and action on these (design team).

Inspection list

Use a defined list of inspections and tests on service systems as part of the re-commissioning process (expert).

Ventilation, heating and hot water systems

Check installation of all equipment, setting of controls and construction details in relation to drawings, specifications, manufacturers' requirements, and equipment manuals and specifications, all as set out in regulatory standards. Test mechanical ventilation systems to compare measured air velocity and flow rates against the designed rate, and check to ensure that all inlet and outlet vents are properly balanced. Test natural ventilation systems in terms of usability and compliance with airflow rates (e.g. all vents, windows and doors) (expert/engineer).

Lighting, water and micro-energy generation systems

Check installation, flow rates and control settings against drawings, specifications, and manufacturers' requirements, and as set out in regulatory standards (expert/engineer).

Monitoring of environmental qualities

The level of physical monitoring needed very much depends on the problems identified and the budget available to investigate them. Spot-measurements using portable four-in-one multi-meters (temperature, humidity, light and sound levels), or short-term monitoring using sensors fixed in key positions, can help to diagnose acute problems. Long-term monitoring is needed to understand complex trends in environmental conditions related to seasonal climate and variation in inhabitant seasonal routines, among other factors. Monitoring and analysis requires suitably trained experts.

In-use monitoring of environment

External environment

An understanding of weather conditions is essential for any BPE study to be able to compare housing performance against design predictions. Cross-check this information against any assumptions made in previous design or compliance calculations. Local weather-station data is good enough for a basic BPE study. For more diagnostic studies, a weather station unit is essential to take account of the microclimate around a home, measuring external temperature, wind speed, solar gain and humidity as a minimum (see Figure 7.6). External measurements of toxicity levels may also be required.

Indoor air quality

Monitoring of CO_2 levels as a proxy for healthy ventilation levels in key areas of the home is particularly useful to undertake during cold periods, when ventilation systems are generally reduced in flow. Measuring indoor toxicity levels related to VOCs, NOx, ozone and formaldehyde, as well as particulates of dust (PM3 and PM10) or allergens, is more complicated and involves sophisticated techniques and equipment.[17]

Temperature and humidity

The level of humidity greatly affects how we perceive temperature. Sensors that combine both measurements are ideal, and are relatively inexpensive to use for basic monitoring. For forensic work, it may also be necessary to measure air velocity using either a fixed or portable anemometer, or to investigate wet-bulb radiant temperatures related to possible heat-stress conditions, using separate sensors.

Thermal comfort survey

This survey is particularly useful to establish what level of comfort inhabitants are experiencing when directly compared to external and internal temperature, humidity and air-speed/flow conditions. It is generally carried out over a relatively short time, as the survey is quite intensive for inhabitants. More details on how to carry out an adaptive thermal comfort survey can be found elsewhere[18] (expert).

Lighting

Spot-measurements using a multi-meter can help to diagnose initial problem areas in a light-touch BPE survey (design team). Compact sensors that combine lighting-level measurements with temperature and humidity are ideal to use for longer monitoring periods (expert).

Acoustics

It may be useful to carry out spot-measurements of noise levels in homes using a simple multi-meter, if any complaints have been made (design team). Further airborne, impact and reverberation testing maybe required to forensically diagnose any noise issues (expert).

Figure 7.6 An installed weather station can provide more accurate microclimate information for BPE

Occupancy patterns

Passive infrared (PIR) motion sensors fixed above door openings reveal how inhabitants are moving from room to room, or in and out of the home. This can help with understanding how and why heating, lighting and ventilation systems are/are not used (expert).

Window and door opening

Binary sensors can measure open/shut positions for doors and windows to assist with understanding environmental conditions. However, more refined sensors are needed for forensic investigation that captures the degree to which windows and doors are opened (expert).

Surveys, interviews and focus groups

Surveys and interviews gather information on what inhabitants know about their homes, which is often far more than the designers or housing managers know. Any interviews should aim to interrogate the key findings of the initial inhabitant survey, and capture the deeper views of the inhabitants and key design and construction team members as well as client representatives related to housing development and management. This is in order to compare design and management assumptions and intentions with performance and inhabitant engagement. Focus groups or facilitated discussion groups can be invaluable for capturing group insights into performance and key areas of significance in the home or related to the neighbourhood, identified in the initial survey and/or interviews. All these methods need careful consideration of ethical requirements before implementing them (see Chapter 13). All these activities need to be independently undertaken by suitably trained personnel.

Surveys

Carry out a basic inhabitant survey to quickly identify any performance issues. Ideally all inhabitants in a home should be surveyed to capture the different perspectives that exist within a household. This can be difficult, and generally BPE surveys consist of just one questionnaire per household. Survey all households, aiming for a return rate of around 20% as a minimum for large developments (typically over 500 homes), and a much higher return rate for smaller developments (typically 1–200 homes). It is quite possible to obtain a 50–100% return rate if the survey team is determined, skilful and experienced (design team or experts).

Surveys with fewer than 100 valid questionnaire returns are not statistically meaningful. However, they are exceptionally helpful in revealing significant qualitative issues. Vital benchmarking is also achievable using the same questionnaire for each housing development evaluated. Carrying out a pre-survey before a retrofit housing project starts can inform the design intentions[19] and enable the benchmarking of a post-occupancy survey carried out after a year or two of inhabitation, which can also help to fine-tune performance.

Care needs to be taken, as people tend to remember the highlights of their thermal experience over a period of time, rather than the overall experience.[20] It is therefore important to carry out surveys in the winter *and* in the summer to capture the significant seasonal variations which affect housing.

Questionnaires

The proprietary two-page Building Use Survey (BUS) questionnaire, originally developed by Adrian Leaman, has formed a cornerstone for many housing BPE studies in the UK and elsewhere, and has been mandatory for several major UK government-funded BPE programmes. It has been refined over several decades and more recently adapted for domestic use. There is now a BUS user group in the UK[21] to help develop this product and share best practice. The BUS survey has an international dataset for benchmarking. Other standard and benchmarked questionnaires have been developed by research institutes.[22] Generally, a questionnaire should not be more than two or three pages long or take more than about 15 minutes to complete, for a high return rate (design team/experts).

Interviews

The semi-structured interview is generally preferred for housing BPE studies, because it ensures that key areas are covered in 10–20 pre-defined questions, while leaving room for the interviewee to add their own insights in a free commentary section. It is best to keep an interview to 30–45 minutes, as attention can drift after this time. Interviewing should ideally take place after the questionnaire, and at the end of any monitoring period, to capture everything. The interview questions should focus on areas of significance highlighted by the questionnaire analysis. Interviews should take place in the home so that the interviewee is prompted by their home environment when answering questions and can make observations in passing (expert).

For a typical housing development, chose a representative demographic spread of households that reflects the general national demographic trend at the time of the study in relation to the housing typology. The selection of representatives from the design and construction teams and the housing developer should reflect the key roles in the housing process. Identifying a good sample of interviewees is critical.

Interviews should be audio recorded, transcribed verbatim and carefully analysed using proprietary software for coding the results into themes. Relying on notes taken during an interview will inevitably lead to subjective bias, as the interviewer will tend to prioritise statements that matter to them, rather than the inhabitant, and will also miss out crucial statements.

Transcription is time-consuming, too – a 45-minute interview can take six hours to transcribe. Speech-to-text recording software can shorten this time, but the evaluator must still check the transcript against the recording for any errors. Coding can take just as long. For this reason, any planned interviews need to be carefully budgeted for from the outset in the project brief for BPE.

Focus groups

Focus groups can act like a group interview. For maximum effect, the number of participants should be between 6 and 12 people. Any more than this will make it difficult for all voices to be heard and for the discussion to focus on key areas. The facilitator needs to elicit all views, not just those from the most vocal participants. Record, transcribe and code the discussion in the same way as for the interviews. A focus group guide must be developed, to cover the key areas of housing performance to be evaluated (expert).

Ethnography

Ethnography is increasingly used in housing BPE evaluation. It involves detailed observation of what is happening in the home, and is useful at all levels of investigation.[23] Generally forensic investigation of this kind is undertaken by academic researchers, but a very basic level of ethnography can be undertaken as part of the home tour by the BPE evaluator.

Observation

Key BPE observations can be made during an initial tour of the home and its immediate surroundings. For a simple home tour, it is important to use a prepared guide sheet that notes whether spaces are inhabited, and if equipment is used as intended. Photography combined with short notes or audio recordings and sketches can quickly capture any issues (design team).

Diaries

For a forensic understanding of how inhabitants are engaging with their homes, it can be useful to ask them to keep a running diary of events and observations. These diaries can then be compared to any unexplained events signalled by physical monitoring to help understand them (expert).

Routines

Deep understanding of how inhabitants engage with their homes can require careful logging of the sequencing of domestic routines, in order to link different processes together which are influencing the physical performance of the home as an intertwined activity. An example could be a cooking routine, which involves the complex use of a cooker, cooker hood, windows, water, energy, task lighting, digital equipment, and other cooking equipment (expert).

Videos

Finally, video methods and techniques are invaluable as a diagnostic tool for understanding housing performance. Short videos taken during the home tour can serve to illustrate an issue in terms of functionality far better than a photo[24] (design team/expert).

All of the above methods provide helpful ways to undertake housing BPE, which needs to be understood within a framework of continuously changing governance, contexts and practices. The next chapter explores some emergent practices in housing BPE, which promise greater insight and impact.

EIGHT
INNOVATION IN
OCCUPANCY FEEDBACK

"I am a camera, with its shutter open, quite passive, recording, not thinking."

Christopher Isherwood,
The Berlin Stories: The Last of Mr Norris
and *Goodbye to Berlin*

Although there is a wealth of existing housing BPE techniques already available, the field is constantly evolving as new insights are revealed, which in turn leads to the need for new approaches to feedback. This chapter explores new directions in housing occupancy feedback, related to prototyping and action research based on the author's own experience. It also discusses emerging techniques that can enhance housing BPE studies. These show how the physical element of 'technologies and products' in the home is interrelated with the understanding of 'institutionalised knowledge and explicit rules', inhabitants' 'know-how and habits' as well as 'engagements',[1] all as 'bundled' practices related to the home and housing in general.

The following aspects are covered in this chapter:

- Using occupancy feedback for product development
- Anthropology, ethnography and action research
- Future directions

Using occupancy feedback for product development

Prototyping housing and BPE

In 2003 the UK Barker Review on housing identified the need to improve supply and quality and lower costs more quickly; the government and the house building industry rapidly promoted Modern Methods of Construction (MMC) as a solution.[2]

Testing inhabited prototypes as building performance evaluation (BPE) for new standard house types can reduce risks and costs, as well as improving chances of success in several ways:

- Highlighting design and construction issues that cannot be anticipated by modelling and tacit knowledge
- Feeding back recommended changes directly into the product development stage
- Refining design customisation and induction processes to take account of user needs and enable 'interactive adaptivity'
- Reducing future costs associated with unanticipated performance issues

As a national house builder, the Stewart Milne Group launched its new Sigma Home standardised MMC volume product through one of 12 innovative demonstrator homes built in the pioneering BRE Innovation Park in Watford in 2007 (see Figure 8.1). The aim was to highlight new MMC technologies and products in the prefabricated 'net zero-carbon' home, and physically test their performance over time. I suggested they had real people living in the home demonstrating a prototype 'in use' rather than relying on computer simulations and physical monitoring alone. We agreed to have a carefully selected family of four to live in the home for four two-week periods spread out over four seasons in 2008.

The living arrangement was not typical, as the family were effectively living part-time in an exhibition site. Nevertheless, this intensive 'prototyping' BPE project provided a wealth of information, involving 30 different BPE methods (see Chapter 12 for methods and cost of study). Four methods were completely new to UK housing BPE studies; evaluation of the inhabitant induction process and home user guide, videoing by the inhabitants of their self-identified design challenges, and logging by the inhabitants of exceptional incidents in the home life based on an agreed template of their routine occupancy patterns.[3]

Figure 8.1 The Sigma Home exhibited at the BRE Innovation Park

An initial co-heating test combined with diagnostic thermal imaging immediately revealed that the fabric heat loss was nearly 50% higher than predicted due to poor and overly complex detailing and construction – an immediate major learning point for all concerned.

Having inhabitants as active research participants paid significant dividends for further understanding the overall product performance. Only five out of the 21 window openings were ever used, showing that fewer openings were needed. The inhabitants developed a sophisticated cross-ventilation strategy to overcome the internal overheating problems generated by a poorly installed heating and hot water system, and poorly designed heating and MVHR controls. They also identified crucial uninsulated pipework 'heat spots' behind the walls and in the store cupboard housing the MVHR unit. The findings led the house builder to completely rethink and simplify the design of the home, leading to the successful new Sigma II Home.

It is important to test new house types and products in authentic, life-like conditions as far as possible before production, to avoid costly mistakes and remedial action. It is possible to reduce the occupancy period to just a week in the winter and summer seasons to capture the full performance range, but any less will not capture the weekly routine of home living or seasonal variations, which are strong indicators related to energy use. It's now also possible to contain a complete house within a two-storey controlled chamber in the UK and test it under different climate conditions.[4]

Using BPE in conjunction with prototype homes also deepens practice knowledge and understanding, as Richard Partington explains:

'What we got most out of the Temple Avenue prototypes was just a much better understanding of building physics – the relationship of the prediction models with the actual detail of design. That gives you the tools to design with. I don't understand why architects give up on it. When I do a design review, I never see architects presenting "So this is the orientation of the building". They never do it.'

Prototyping products and BPE

Based on the success of the Sigma Home BPE project, the author joined a consortium to help develop a £6.4-million housing research and development project known as AIMC4 (Application of Innovative Materials, Products and Processes to meet the Code for Sustainable Homes Level 4 Energy Performance). The aim was to pioneer and deliver low-carbon homes through innovative fabric and building services solutions, testing them through BPE.[5] We used hands-on focus groups to explore the usability of the key technologies proposed – something quite common in digital product development, but not widely applied to heating and ventilation products at that time. Existing homeowners drawn from a recent project developed by one of the participating house builders were encouraged to individually use the sample windows and MVHR units mocked up to reflect real-time conditions. They then discussed, within a group setting, what they experienced individually when trying to use the technologies.

The results were startling. In one case, a well-known window product was found to be completely unintuitive in terms of how to open it first time, as it had three different levers to reflect different types of opening conditions - it should be easier (see Figure 8.2). In another case, the heating zone control interface was unintelligible to users.

Figure 8.2 Windows should be simple to operate – a single lever with a simple action

Perhaps the most interesting finding was that that only 14% of the window product suppliers contacted actually tested their products on real humans for usability.

Usability survey

Relatively little is known about the interface between inhabitant and home controls, which can be quite baffling. A BPE usability tool has now been developed, which consists of a matrix with a five-point assessment scale for evaluating these products against each of the following criteria:

- Clarity of purpose
- Intuitive switching
- Labelling and annotation
- Ease of use
- Indication of system response
- Degree of fine control

The products are evaluated either by an expert or by the inhabitants themselves, working with an expert.[6] The 'touchpoint' controls evaluated are heating, ventilation, lighting, water, cooking, and security fixtures and fittings. Figure 8.3 shows a typical evaluation matrix.

The usability tool works best when combined with a home tour, inhabitant interviews, induction evaluation and document review, which help to reveal the underlying causes of usability issues. The usability of the MVHR unit shown here was compromised by the position of the door frame blocking the air filter access, and the high position of the unit itself.

Description and location				
Greenwood Filters within MVHR unit				
Usability criteria	**Poor**			**Excellent**
Clarity of purpose				
Intuitive switching				
Labelling and annotation				
Ease of use				
Indication of system response				
Degree of fine control				

Comments

No labelling whatsoever – if you are not familiar with the unit you don't know they are there. The filters are easy to take out and clean but some units clash with the door frames. There is little indication of when they are getting clogged up – a warning light might be useful? The whole front panel must come out to access the heat exchanger filter every two years – this operation in some units on site is very difficult due to clashes with doorframes and very poor location of units high up in narrow spaces.

Figure 8.3 Evaluation from the usability tool of an MVHR unit

However, understanding isolated individual technologies can never be a substitute for considering how all of them perform together within the totality of the home as it is lived in. An anthropological approach can enhance traditional housing BPE methods in this respect.

Anthropology, ethnography and action research

Learning with the inhabitants

Social anthropology operates primarily through participant observation, where the aim is '… to notice what people are saying and doing, to watch and listen, and to respond in your own practice.'[7] This turns housing BPE into attentive learning *with* inhabitants, rather than a study of inhabitants' behaviour in order to influence them, and it has powerful implications. It is quite different from the 'expert knows best' approach traditionally adopted in BPE. The anthropological approach acknowledges that inhabitants often know more than evaluators about the performance of their homes, as lived in products. The outcome of the study is always uncertain, and demands that evaluators open themselves up to reconfiguring and reframing their understanding and approaches to housing BPE methodology, depending on what they learn with others.[8]

One example of this 'reframing' involved the BPE team having to re-design the usability tool. This arose from participant observation and subsequent discussions with the inhabitants of a newly retrofitted housing block in Leeds. It became obvious that the tool did not work on its own – the evaluator was needed on hand to help explain the tool and understand the context.[9] In the same project, the inhabitants themselves introduced us to another new method for BPE. They were using a social media site to discuss housing performance issues together and solving them collectively, as discussed later in this chapter.[10]

Action research

Participant observation becomes action research when the BPE evaluator engages directly with the inhabitants and co-learns about various issues in the home through dialogue. This helps inhabitants to overcome the issues raised, by providing expert guidance and raising the issue with the housing client and design team.[11]

In the Leeds retrofit project mentioned above, the inhabitants highlighted poorly designed windows and ventilation controls, which the evaluators then raised with the housing developer. Over 90% of the 95 inhabitants surveyed did not use their ventilation system as recommended, either because they were not aware of it, they did not know how to use it, or they turned it off because it was too noisy. This was one cause of significant overheating in the apartments. Many inhabitants were using their internal blinds to adjust lighting quality, rather than as shading to prevent overheating. The evaluators then explained to the inhabitants how they could improve their comfort through appropriate shading and ventilation practices based on the BPE findings, as a form of action research.[12]

Visual ethnography

Ethnography is an account of life as it is actually lived and experienced by people in a particular place at a particular time. As a study, it can produce recommendations for interventions in any given housing BPE process, or speculate on possible future changes for housing design based on an understanding of hidden rules, know-how and habits operating in the home.[13] The use of video as a visual ethnographic method is often more powerful than a photographic survey in revealing performance issues and causes. Video can contain a voiceover as narrative, and can show how people relate to their own homes, through conversation with the evaluator.[14] It can also be used to examine any product through a carefully designed experiment on site to understand functional performance issues and help improve the design.[15] Who is doing the videoing is important, as this can clearly change how any video 'represents' the home as an interpretation of reality.[16]

An important innovation in the Sigma Home project was the development of a short 'video tour' in order to understand more about issues arising in the home.[17] This gave a new sense of agency and voice to the inhabitants. The work was based on an earlier method developed by Sarah Pink in 1999.[18] The inhabitants were asked to video their own experiences of living in the home, without the presence of a researcher. This removed any researcher bias and was far more cost-effective. They were encouraged to record issues selectively in terms of importance, using just one hour of video time during each occupancy period. They were also asked to provide vocal descriptions of the issue, as they were videoing it, to aid the evaluator's understanding.

The video footage was then coded against key functionality factors to identify any issues.[19] This allowed numerous unforeseen issues related to the intended performance of the home to feed back into the design process. One example was the capture of the mother's attempts to cook a meal, only for the smoke alarm to go off each time, because it was positioned far too close to the cooker. This resulted in her opening the windows wide for a long period to stop the smoke alarm, leading to an uncontrollable loss of heat in the winter.

Care must be taken in terms of ethical procedures to avoid any violation of privacy for the individuals concerned. It is important to discuss the video with participants, and anything they do not wish to be publicly available must be kept anonymous or deleted.

Social media and BPE

Another innovative way to understand and learn about housing performance is through social media, as electronic communication through which users create online communities to share information, ideas, personal messages and other content. About 77% of online UK adults claim to have a social networking profile.[20] It is a powerful medium for people, who don't know each other and who live in housing developments with the same type of construction and installed technologies, to learn as a 'digital community' how to improve their living conditions. This can add a powerful dimension to occupancy feedback for design improvement through a simple harvesting and analysis of existing commentaries.[21]

In the Leeds retrofit BPE project, the evaluator discovered that the inhabitants had set up a very lively closed Facebook group with 466 members and an 81% engagement rate within the period analysed, which the evaluator subsequently asked to join and have permission to analyse. Qualitative analysis of the Facebook posts revealed numerous performance issues compared by the members, with answers provided using pictures and even purpose-made DIY videos or offers of direct help. The key areas concerned temperature issues, noise and lowering the cost of living. In many cases, the exchanges were so positive that members ended up sharing personal details in order to help each other more. Advice included installing curtains and keeping them closed, closing air vents and door gaps, wearing warmer clothes, changing energy provider, and obtaining more efficient heating equipment than the thermal storage heaters provided. To be effective, these type of groups depend on knowledgeable members who can support others and prevent the spread of misinformation, particularly in relation to less tangible issues such as indoor air quality.

The analysis of members' interactions nevertheless provided a great deal of extra insight as to what was happening in the homes which affected their performance.[22] The evaluator provided guidance based on BPE findings as they emerged, which was then informally disseminated via the inhabitants' Facebook group. Inhabitants were thus empowered to discuss their own approaches to solve new issues based on the findings. The powerful combination of a BPE project with social media discussion helped to deeply embed the project findings as well as disseminating them to a much wider audience. The dissemination of know-how occurred strictly on a 'need to know' basis, which was much more effective than simply disseminating the findings.

Future directions

Open-source digital media

While digital 'big data' information gathering and the 'Internet of Things' tantalisingly offer enhanced BPE understanding and practice, they can be more empowering for the inhabitant when the knowledge gained is shared openly. This increasing demand for open-source data has led to new initiatives empowering citizens to become 'scientists' who provide mass data to open-source research projects. Citizens have started to take collective action to monitor air pollution, highlighting where more attention needs to be given to overlooked health issues. Such monitoring can open up possibilities for changing the very conditions that are being monitored. Portable DIY monitoring techniques can provide a more detailed picture of an individual's exposure, and include their own sensory experience alongside the monitoring data. Taken collectively, these readings and experiential accounts can build up a much richer context of understanding, accepting that they provide 'good enough data' rather than absolute accuracy.[23] In housing BPE, inhabitants can easily become involved as citizen scientists to augment the BPE project, and also to contribute to the definition of what should be monitored, given their experiential knowledge of their own homes (see Figure 8.4).

It is also possible to 'data mine' open-source BPE projects, using big data to undertake meta-studies related to general housing performance.[24] These meta-studies can examine specific themes such as summer overheating,[25] or simply co-verify case study results from across different projects.[26] Concurrently, the advent of smart homes associated with the internet of things allows monitoring of technology use at a far more detailed level and across a far wider set of home living practices than previously possible. However, these smart technologies can be disruptive and unfamiliar to the inhabitant. They may harvest personal information and use data for third parties – a potential invasion of privacy. They can also demand that the inhabitant adapt without necessarily providing any energy savings.[27] Future housing BPE needs to reflexively question the need for any new technology in the home, on a continual basis.

Participatory action research

Action research with inhabitants in housing BPE could be used in temporarily inhabited 'show homes', or prototypes ahead of the production, to routinely test new homes. A major step forward would be to include the inhabitants as co-evaluators, who are able to co-design the BPE criteria with the professional team. They could also have input into the analysis as a form of 'conversation' to help articulate tacit BPE findings that remain inexplicable to the evaluator and the design team. This participatory action research could improve the design of new housing typologies by reflecting both improved criteria and understanding of behaviours and practices, as Richard Partington discovered in his Derwenthorpe housing project:

'… because the family moved in and because we had conversations with their occupational therapist, you got a bit of an understanding about what an adaptable home needs to be like – things like integrating a platform lift, adapting kitchens, evaluating whether there was enough storage, whether the controls were actually usable by all of the members of the family.'

Co-housing and other collective self-build developments as new housing typologies demand that BPE evaluators engage with the housing community as well as individuals, due to the collective involvement of inhabitants in the design and maintenance of their housing development.[28] Discussing the results of initial and later BPE findings with the community develops a virtuous circle of learning. This type of collective participatory action research can become part of a BPE process to inform future design work undertaken by the same community, or other communities looking at a similar housing typology.

A good example of this type of action research occurred when the research team worked closely with the LILAC co-housing community in Leeds to explore how well the homes were performing, and help the community improve the performance of their homes. There was regular interaction between the research team and the inhabitants through emails, visits

Figure 8.4 Inhabitants can become 'citizen scientists' to help with BPE studies

and group meetings. The collective reflection of the community helped them to develop priorities for remedial action. This included revising the guidance for handover procedures and dealing collectively with the emerging issues concerning the installation of the MVHR systems. The learning process resulted in action to recommission the MVHR system in all dwellings, insulate ducting, label MVHR boost switches, open dampers in MVHR fan units that remained closed (preventing proper functioning) and instruct all inhabitants to use the cooker hood function when cooking if they switched their MVHR system off. LILAC also developed a community maintenance system for all their MVHR filters.[29]

Institutional innovation

This chapter would not be complete without a brief consideration of how institutional innovation can assist the development of housing feedback. Firstly it is clear, as a result of the tragic Grenfell Tower fire and the subsequent government Hackitt Review, that legislation and regulation needs to be more focused on the safety and wellbeing of inhabitants over the entire life of a building. BPE has a vital role to play here in terms of providing evidence of performance in reality, and should be incorporated into this legislation as a first step, and subsequently into wider domestic building regulations. Secondly, all the built environment professions should respond to the Hackitt Review by mandating BPE as part of 'closing the loop' and becoming champions for inhabitants, rather than champions just for themselves. This begins with education, as examined next.

NINE
EDUCATING FOR FEEDBACK AND LEARNING

"I did then what I knew how to do. Now that I know better, I do better.**"**

Maya Angelou

Why is there so little BPE taught in education and in practice? Part of the answer lies in architectural education that is forward-looking, aiming to judge performance based on design and prediction, without any direct feedback from reality. The performance gap has only recently been raised as an issue for architects, and there is relatively little engagement with understanding the inhabitant. As such, students are largely asked to develop fantasy projects based on 'real life' briefs and sites, but with no built output to reflect back on, no inhabitant to talk to, and no means of developing the accuracy of their predicted performance beyond simulation. As a result, learning from the next project becomes important, not from what happened in the last one, and so it goes on into practice – as a bad habit.

Fortunately, this is changing. New forms of graduate apprenticeship routes are opening up, offering excellent opportunities for students to learn about BPE in practice, and break this habit. Students on architecture and engineering programmes are also increasingly engaging with 'live projects' that involve design and build, again offering a starting point for effective BPE feedback to inform their future designs. At the same time, national NGOs and professional institutions are recognising the need to provide training in BPE as part of normal CPD packages for their members.

Five groups benefit from BPE education: designers, to avoid past mistakes; educators, to pass knowledge on to students; inhabitants, to get the best out of their homes; housing developers and managers, to deliver and manage a good product; and, by implication, policymakers, to improve housing design and performance. This chapter explores methods for teaching BPE and the training opportunities provided in the UK. It concludes by describing the personal attributes and skills that a practitioner needs to undertake housing BPE studies.

The following aspects are covered in this chapter:
- Educating the designers
- Educating the educators
- Educating the clients
- Educating the inhabitants
- What makes a good housing feedback practitioner?

Educating the designers

Knowledge is embedded through people and communication. One of the best ways for professional practitioners to learn about BPE, therefore, is through knowledge exchange (KE) between different communities of practice (CoP).[1] Community members are bound together through mutual engagement to develop a shared repertoire of communal resources over time.[2] CoPs can operate within an individual practice, or across many practices.[3] The architectural Practice Leads network in the UK is an example of a national CoP supporting research in practice. The development of BPE KE is most effective when human 'brokers', such as researchers, consultants or other BPE champions within practice, can translate, coordinate and align perspectives between these CoPs using various tools and techniques. These brokers also need to have a strong enough reputation to be able to influence the development of the staff within a practice, and introduce elements of practice to each other to enable learning to take place.[4]

Embedding effective BPE through organisational learning can help professional practices to:

- improve their performance (single-loop learning)
- question their assumptions (double-loop learning)
- open up intuitive new ways of being, knowing and doing (triple-loop learning) (see Figures 9.1 and 9.2).

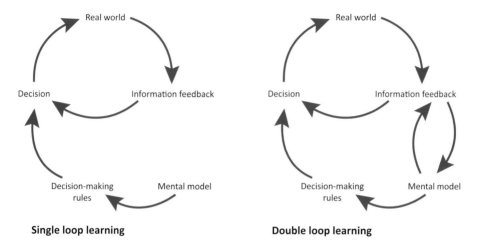

Figure 9.1 Single- and double-loop learning in individuals

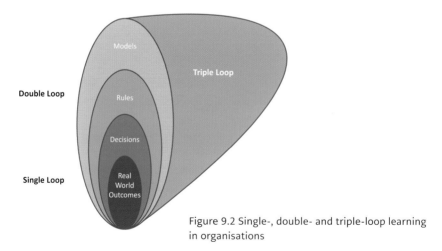

Figure 9.2 Single-, double- and triple-loop learning in organisations

BPE practice champions ideally need to lead from the top, working directly with practice directors or CEOs to obtain maximum leverage.[5]

One model of BPE KE enabled a graduate associate engineer to act as a BPE 'broker' in a two-year Innovate UK Knowledge Transfer Partnership with a leading practice, Architype Architects. She developed ten iterative BPE studies of Architype's buildings, working in collaboration with others, and with support from another academic, the practice directors and me. She gradually built up credibility and trust, challenging the practice to question tacit design assumptions and work procedures in order to improve their building performance. The practice developed new tools and processes for deeply embedding BPE learning in every aspect of their approach to producing architecture. This included an internal knowledge-sharing website using an open-source mediawiki framework with multiple plug-ins, as well as internal continuing professional development presentations.[6] BPE is now part of Architype Architects' DNA as a CoP; they feed forward findings from their post-occupancy studies to improve all their design work and build in BPE costs from the outset of any project[7] (see Figure 9.3).

Figure 9.3 Architype's Enterprise Centre, East Anglia University

A different model of BPE 'brokering' occurred when a well-known international practice, Buro Happold Consulting Engineers, collaborated with academics to develop a unique industry-based engineering doctorate programme. One PhD student on this programme completed an in-depth post-occupancy evaluation (POE) study focused on the impact of occupant behaviour on the real-life performance of aspiring low-energy/low-carbon housing in the UK.[8] Buro Happold learned significant BPE lessons through this particular 'broker', who went on to provide similar services to one of the largest construction companies in the UK.

A third BPE KE model is a service company that embeds interdisciplinary consultants within client organisations for a period of time, tying in with the latest British Standard Code of practice for facilities management, which requires POE as part of the initial briefing for any project and then annually for three years after handover.[9] This model can effectively transcend the different disciplinary approaches towards BPE from architecture (qualitative and improvisational)[10] and engineering (quantitative and replicable),[11] combining the best of both.

All three models described here offer different ways to embed BPE in practice (see Figure 9.4).

Housing Fit For Purpose: Performance, feedback and learning

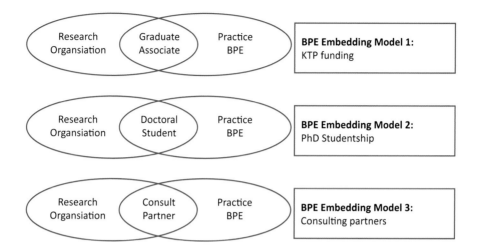

Figure 9.4 BPE deep-learning models for practice

The RIBA has developed a POE portal with basic guidance for design professionals to use,[12][13] and is planning to develop systematic training in BPE for its members. For professional practices to understand the real value of BPE in developmental and financial terms, training has to be deeply embedded within a design practice using 'champions' and 'brokers' who stay around long enough to ensure that a genuine culture change takes place from within. Unless BPE becomes an integrated and iterative practice-based learning process, it can be easily lost when a 'broker' leaves. There also needs to be a stronger educational mandate for BPE from all built environment professional bodies. This includes surveyors and valuers, who have an important influence on whether or not BPE is included within contracts.

Educating the educators

An introduction to BPE methodology and practice within the first few years of architectural and engineering curricula gives students a realistic and deep understanding of how buildings perform, how they are used, the concerns that arise from inhabitants, and how to test their design assumptions/strategies. It should be framed as a routine activity that is core to the design process in the design studio, and not as a bolt-on activity. Other built environment education programmes such as planning, surveying, real-estate and construction should also include an element on BPE to support all built environment processes.

Initiatives in the USA

Gary Moore taught POE to postgraduate architecture students as early as 1975 in the USA, based on a single-semester course, during which students carried out extensive social surveys and environment behaviour analysis.[14] The national Vital Signs education project (1992–8) followed on, with an increased focus on understanding environmental performance. Students physically monitored and evaluated the performance of existing buildings as case studies, and were encouraged to apply their new knowledge to their own design work.[15] The follow-on Agents of Change project delivered training to over 50 architecture education programmes, and annual intensive POE workshops for practitioners known as Tool Days.[16] POE is now a required part of the live architectural design and build curriculum for projects such as Ecomod in the University of Virginia,[17] and the bi-annual Solar Decathalon international competitions.[18] The Society of Building Science Educators (SBSE), an open international platform hosted in the USA, has been promoting POE methods in teaching since its inception in 1995. There have also been local research-led teaching projects using POE methods in other countries, as evidenced in numerous passive and low-energy architecture international conference presentations since 2005, often based in Master's-level programmes.

Initiatives in the UK

Following early teaching by Tom Markus at the University of Strathclyde in the 1970s and 1980s, BPE methodology and practice has been taught at postgraduate level by enthusiasts in some higher education establishments in the UK, such as Oxford Brookes University.[19] As part of a team there, I taught POE in the second and third year of the architectural programme, inspired by the USA Vital Signs project. In 2010, we also set up a one-week project called Decarbonised Desires, to provide all our fifth-year students with hands-on BPE group work to inform a design intervention.

The seven steps in the Decarbonising Desires project were as follows:

1. Identify an existing benchmark standard for CO_2 emissions. Choose your own benchmark level with a justification.

2. Survey an existing building: plans, section, elevations (obtained from the planning department and building control). Do a one-hour walkthrough to identify any issues of ventilation, heating and lighting leading to higher carbon emissions than expected.

3. Find out actual fuel bills and running costs from the facilities manager for the whole building. Identify the worst-performing aspect of the building from data obtained, and select a part of the building to focus on.

4. Calculate U-values for the chosen part of the building.

5. Model this part of building using dynamic simulation (a room or group of rooms) and analyse heat loss/gain, ventilation rate, light quality (radiance) and levels.

6. Design a physical intervention to this part of the building that could improve performance towards the desired carbon emissions benchmark.

7. Test the intervention using dynamic simulation.

The project ended with a successful public exhibition in Oxford and an open review of the proposals, with citizens contributing their views alongside an expert panel.

We also initiated a series of three national workshops in 2008 for all schools of architecture called Designs on the Planet that revisited the technology curriculum to improve building performance. Thirty-four out of the 41 schools of architecture in the UK participated.[20] However, there was little subsequent development of BPE as a mainstream undergraduate activity in these schools.

Puzzled by the lack of progress, I led a team to develop a national conference in 2015, 'Beyond Building Performance: Architectural research, practice and education' to explore the issue further.[21] Representatives attended from virtually all schools of architecture in the UK and promptly sent a manifesto statement to the RIBA Education Committee:

'This conference believes that integrating BPE within education is essential in order to: fulfil our responsibility to society, exploit the potential for collaboration between academia, users, research, disciplines and professional practice in expanding the evidence base for affordable, biodiverse rich, healthy, resource efficient building and urban design, supportive of communities.'

To date, however, the ARB and RIBA validation criteria from programmes in architecture do not explicitly require BPE competencies in graduates. There are only overarching statements about understanding '… the needs and aspirations of building users' (criterion GC5.1) and '… the impact of buildings on the environment, and the precepts of sustainable design' (criterion GC5.2).[22] Until BPE is a validation requirement, the number of schools of architecture teaching BPE routinely to all students will probably remain small.

By contrast, the UK Royal Academy of Engineering's Centres for Excellence in Sustainable Design initiative has led to interdisciplinary courses which teach evidence-based design related to building physics and performance evaluation.[23] A joint institutional initiative like this could bring disaggregated efforts by professional institutes more closely together in relation to embedding BPE in higher education and practice. Understanding how institutional and political considerations affect knowledge perception and evaluation of the value of BPE across different types of audience is also vital in this respect.[24]

Educating the clients

Housing developers could also benefit from the same type of education in BPE as design professionals, using CoPs and KE across the sector. There is, however, no institutional housing forum promoting BPE training at present in the UK. Instead, training devolves to private housing developers through national organisations such as the National House Building Council (NHBC), the Chartered Institute of Housing (CIH) and the Home Builders Federation (HBF). Presently, none of these organisations have training policies that include BPE as a standard activity. The first step here is to create an understanding of the value of BPE with the leadership and members of these organisations to be able to develop policies mandating BPE as a routine activity in the housing sector.

One excellent CoP that has consistently promoted BPE is the Good Homes Alliance (GHA). This national organisation undertook in-depth POE activities on several innovative housing case studies in England, in a project funded by government agencies and the NHBC, working with several academic partners from 2011–13. Recommendations were fed back the members and made publicly available via a pioneering BPE report[25] (see Figure 9.5). The GHA have regular training events related to housing performance and publish other POE reports on their website, demonstrating excellent knowledge transfer. Other CoPs such as the UK Green Building Council and the Association for Environment Conscious Building also promote POE to their members through seminars and conferences. Some housing developers have successfully trained themselves to access BPE via government-funded projects such as AIMC4,[26] as this makes such activity more affordable.

The National Housing Federation and the government executive agency Homes England also have a role to play in promoting BPE training and policymaking. While the former has not yet actively promoted such training, the latter was poised to embed BPE as the first-ever government Soft Landings policy and process for housing through a national flagship zero-carbon housing programme in 2009[27] – sadly put on hold due a financial recession and a change of government.

There is a rapidly growing community-led housing sector with interest from government and other funders in supporting this as an alternative to mainstream housing. The BPE champions are usually members of the community group who have the relevant technical expertise, as discovered in the two-year BPE project we undertook with the LILAC co-housing group in Leeds.[28] An effective client representative is crucial – someone who can objectively transmit any BPE findings back to the community group and design team, to

Figure 9.5 Innovative housing in Stawell, Somerset, assessed by the GHA using BPE

aid their learning and represent the overall client view during the whole BPE process. BPE training for this sector is also needed.

One final client worth mentioning is the 'client's client' – the financial lenders and insurance companies. They have a huge influence in the housing sector, determining what is perceived as valuable within housing development. The NHBC warranty, for example, still has no requirement for BPE to be undertaken when a housing project is complete which can mitigate risk and costs. These institutions also need educating to understand just how much BPE can do for them and their members.

Educating the inhabitants

Educating inhabitants in relation to BPE is not simply about teaching them to behave appropriately and 'simplifying' control settings for them.[29] That is pointless if it is the design, construction or installation causing the problem in the first place. It is far better to empower inhabitants through education on how their home works and performs in real life, so that they can respond to any situation with greater knowledge. This involves understanding controls, knowing how to optimise the system, and knowing if it is malfunctioning or not. BPE can be exceptionally powerful when used as a learning tool for inhabitants in this way. Clients often do not report housing BPE findings to the inhabitants through fear of an adverse reaction or not believing that there is any value in doing so. This is a missed opportunity, as inhabitants are often in the best position to tell housing providers *why* something is not working, and not just *what* is not working. Focus groups are good for discussing BPE findings with inhabitants, providing there is some expectation management. Housing satisfaction surveys miss out much of the vital performance feedback that broader BPE studies provide.

BPE also provides an excellent opportunity for developing inhabitants' routine engagement with the performance of their homes. Feedback can indicate whether they need more ventilation, and whether their heating and cooling systems are working well or not. An engaged inhabitant will also be much more likely to maintain these systems. Inhabitants need to have monitored feedback for indoor air quality (particulate/carbon dioxide metering), energy use, water use, air temperatures and humidity levels ideally.

BPE needs to take account of inhabitant demographics for training purposes – reports and presentations need to be accessible and easy to understand for older people as well as for those whose first language is not English. Providing the feedback via a website is tempting, but disenfranchises a large number of elderly people who are not digital natives. Face-to-face meetings are a far more effective means of providing feedback, particularly when a generous question and answer session is provided. Education for inhabitants goes much further than the induction that housing developers provide at handover – it should be an ongoing activity throughout the life of the home, alongside routine BPE activity and when any changes occur in the home, such as new inhabitants moving in, or any upgrading. Whoever introduces these changes (e.g. the landlord, builder, installer or estate agent) should also be responsible for the necessary training.

What makes a good housing feedback practitioner?

A housing feedback practitioner is an 'honest broker' who tells the truth about housing performance, tactfully and palatably, while recognising the limits of the situation they are working in, and the relative nature of that truth. There is no point in disguising negative findings, and every reason to praise the project in relation to positive findings. It is also important to understand what can change in the short, medium and long term, and not alienate the client with unrealistic expectations or recommendations. A practitioner should have the profound desire to understand the aims of the client commissioning the BPE study as well as the aims of the project and the brief resulting from these factors. A broader understanding of the political and legal context of the project is important, to ensure that the feedback attunes to these elements and the client can thus 'hear' what is being reported back. A practitioner must also have a strong and clear set of ethics related to their BPE work (see Chapter 13).

Housing BPE can be undertaken by anyone who has a good understanding of the design and build process, the right attitude and relevant training. BPE is often carried out by a small team, combining qualitative and quantitative skill sets. It can be hard for the project team to be objective about the performance of their own product, and so it is preferable for an independent team to carry out the BPE, or at least for an independent person to lead the BPE team. Members of the project team can still be involved in doing the various tasks, particularly at the 'light-touch' level of BPE, but independent oversight is important. In a large design organisation, a separate design or research team can undertake the work, but there can still be a danger of bias in the management review of the final BPE report to protect the organisation's vested interests. Building up expertise in BPE takes time. A good start is for a new practitioner to accompany an existing BPE expert on site visits in order to learn the art and practice of BPE evaluation. This type of activity is just as important as becoming literate in the building physics that underpins BPE.

The very best practitioner is someone who can combine all of the following personal attributes and skills for a detailed BPE investigation.

Personal attributes

- Empathetic – understanding where your client is coming from, and relating to the inhabitants' situation too
- Sensitive to the home and neighbour context, ethical, objective, curious, problem-solving and solutions-oriented
- Persistent, with the ability to probe fearlessly
- Willing to understand and relate to different contexts and disciplines; a team player
- Able to recognise personal limits and delegate to others with appropriate skills
- A good time manager in relation to multiple and contingent tasks
- Diplomatic and tactful, with a strong desire to communicate with others
- Interested in statistical comparisons, persuasive, and willing to learn each time from the process and revise it accordingly, rather than simply repeat a stock approach

Practical skills

- Understanding of basic building physics, spatial design, construction strategies, processes and detailing
- Able to identify, organise and coordinate the resources necessary for the study
- Able to work with different disciplines and contexts
- Able to organise a case study, and carry out a contract document review, construction audit and an energy and water audit
- Able to design and undertake physical monitoring; to prepare, order, calibrate and install monitoring equipment, observe what is different from expectations during a site visit, undertake a photographic and thermal imaging survey, and take spot measurements on site and in the home
- Able to organise and undertake a home tour with video and/or photographic recording, undertake interviews with the inhabitants, design team, contractor and client representatives, and organise and undertake a questionnaire survey of inhabitants
- Able to organise and store multiple data sources, undertake statistical analysis in relation to appropriate benchmarks, analyse and triangulate data from multiple methods, synthesise findings from multiple sources, and make an evaluation of the findings in relation to the specific context in order to produce viable recommendations for action
- Able to write concise and succinct reports, and prepare dynamic, visual presentations that engage the audience with the findings and recommendations

A person with all these skills is very rare, and so it is more important that the BPE practitioner has the necessary personal attributes and is then able to organise a team who has the necessary skills, alongside their own, according to the level of BPE activity to be undertaken. There is a real opportunity for design professions to engage with the creation of BPE practitioners like this. If they do not, it is likely to become a separate discipline, further diminishing the role and authority of existing designers by breaking the link between evidence and design. Having covered the training needed for BPE, the next section of this book demonstrates examples of successful BPE teams in action, and typical costs.

SECTION

APPLICATION AND CASE STUDIES

4

TEN
THE INTERNATIONAL CONTEXT

"All knowledge of cultural reality, as may be seen, is always knowledge from particular points of view."

Max Weber

This next section of the book evaluates BPE through a series of case studies. This chapter examines three international examples, while Chapter 11 provides a detailed account of two UK BPE projects. Chapter 12 explores the costs associated with BPE studies using cost-benefit analysis to identify a best-value approach.

The international development of BPE in practice has been complex and uneven, due to the individual histories, cultures, societies and environments that deeply inform practice in any given place. BPE methods and standards vary from country to country. International standards vary also. Practitioners need to familiarise themselves with the relevant contexts and standards before undertaking a study in a place which is unfamiliar to them. The international examples highlighted here help to illustrate some of these differences.

The broader significance of any BPE study depends on its immediate objectives, how it helps to develop BPE methodology and techniques, and what impact it has on policy development. Studies of newbuild housing have dominated to date, and the role of BPE in retrofit projects has been relatively underdeveloped, despite the significant impact of upgrading existing housing stock to reduce global carbon emissions. Retrofit case studies in this chapter and the next one highlight crucial differences in approach when compared to newbuild BPE.

The following aspects are covered in this chapter:

- International standards, cultures and knowledge exchange
- International newbuild BPE
- International retrofit guarantee
- Mapping of techniques deployed internationally

International standards, cultures and knowledge exchange

International standards

A number of different voluntary international building design performance standards have been developed in response to the drivers set out in Chapter 2, and have been subsequently compared in terms of the different predictive evaluation methods they use.[1] Other more uniform global standards for building monitoring and reporting, such as the International Performance Measurement and Verification Protocol (IPMVP),[2] tend to be limited to energy and water use only, and are mainly used for large non-domestic projects.[3] Although uniform standards enable international comparison, they can be rigid, slow to change and fail to take account of local conditions and needs.

Two leading sustainability standards with housing design certification are Leadership in Energy and Environmental Design (LEED), launched in 1999 and operating in 165 countries, and Building Research Establishment Environmental Assessment Method (BREEAM), launched in 1990 and operating in 77 countries. Both are market mechanisms for promoting buildings, however, and have only recently moved from predicted performance to a degree of measured performance.

BREEAM has ten areas of predictive evaluation: management, health and wellbeing, energy, transportation, water, materials, waste, land use and ecology, pollution, and innovation. It finally made post-occupancy reporting of CO_2 emissions mandatory for all its standards in 2018, but nothing else. LEED v4.1 Residential has nine areas of predictive evaluation: integrative process, location and transportation, sustainable sites, water efficiency, energy and atmosphere, materials and resources, indoor environmental quality, innovation, and regional priority. Like BREEAM it also demands post-occupancy reporting (for five years minimum), but of energy use rather than carbon emissions, and again, nothing else. Both certification systems offer optional credits for undertaking wider housing POE. LEED also has a set of Global Alternative Compliance Paths to respond to local housing conditions.[4] Significant differences exist between local, national and international building performance evaluation tools. In one Hong Kong case study of two housing developments the authors showed how international standards can reveal shortcomings in local standards.[5]

Ronald Rovers, a consultant heavily involved with the development of building performance evaluation in the Netherlands, is critical of LEED and BREEAM:

'These are all tools with 100 categories and green impact factors, and everything is hidden. I even think that is the aim of the building performance industry, to get these tools expanded so that an individual fact is replaceable. So, these tools won't help us get the right register, and that's a major flaw, because they've got too many parameters, which hides the individual CO_2 emissions. If you want a building performance with lower CO_2, we have to measure CO_2. And on top of the other categories, they apply weighting factors to come to one figure and outcome, which is pretty useless.'

The Living Building Challenge (LBC), launched in 2006, and the WELL Building Standard (WBS), launched in 2014, are two more recent certification schemes. Both demand a more complete POE study after 12 months as mandatory, although the WBS uses spot-check measurements only. Certification for both depends on submission of the POE information to demonstrate compliance. The LBC includes seven key areas for evaluation: site, water, energy, health, materials, equity and beauty. The last two areas are unusual in POE studies, and it is good to see this extra level of development. The WBS has 11 different areas of evaluation with a specific emphasis on healthy buildings and wellbeing: air, water, nourishment, light, movement, thermal comfort, sound, materials, mind, community and innovations.

These four international standards show no consistency in terms of their relationship with BPE, which makes international comparison between them impossible. They have different approaches to defining the areas to be evaluated. These relate to the individual drivers operating in the country in which each standard was first developed.

International knowledge exchange

International knowledge exchange related to housing POE is, however, well developed in the Passivhaus standard, with the performance of Passivhaus housing assessed in various countries[6] and discussed at international conferences. However, this has been more on the basis of proving the performance of these homes rather than promoting BPE for design improvement.[7]

The development of BPE raises a number of cross-cultural challenges. What works in one country or culture does not necessarily work in another. In response to this, I developed a three-year knowledge exchange process with Dr Magda Baborska-Narożny to embed BPE know-how from England in Poland via a championed network, and for English BPE experts to learn about the Polish context.[8] This EU-funded project was called BuPESA

(Building Performance Evaluation for Sustainable Architecture). At the culminating bilateral BPE symposium held in Poland in 2016, policymakers, practitioners and academics were constrained by differing national government viewpoints on the need to reduce carbon emissions, as well as completely different housing market histories.[9] These different understandings and aspirations affect the definition, scope and content of BPE. Nevertheless, the host city of Wrocław is now undertaking its first-ever city-wide housing POE studies in relation to air quality.

International newbuild BPE: Brazil

Background
A further international BPE knowledge exchange project involved collaboration between two Brazilian academics and myself. We studied a large affordable housing development in Brazil and shared innovative BPE methods that each of us had developed over time. This case study of the Shopping Park housing development on the outskirts of the Brazilian city of Uberlandia illustrates how BPE can operate at a neighbourhood level using action research and co-production methods to inform strategies for improved resilience in social housing.[10]

The development was built between 2010 and 2013, and procured 3,000 homes under the problematic national Minha Casa Minha Vida government housing programme in Brazil, which has built three million new homes to date.[11] All the homes were the same Minha Casa Minha Vida house type: single-storey, semi-detached, naturally ventilated, with a living room, two bedrooms, kitchen and shower/toilet room totalling just 33 m^2 of usable floor area (see Figure 10.1). Solar roof panels, supplemented by electricity, provided the hot water. The construction was rendered clay block walls, metal-framed single-glazed windows, and a shallow-pitched ceramic-tiled roof on a timber frame with a PVC panelled ceiling. Each home sat within a garden plot that allowed for extension to the rear and front (see Figure 10.2).

Methods
An overall evaluation of the neighbourhood (3000 homes and associated amenities) in 2016 utilised innovative co-production methods to identify the key issues. Capturing stakeholders' concerns in three mapping workshops was a powerful tool, providing everyone with a chance to stake out their issues and priorities and debate these collectively. The inhabitants clearly suffered from a poor environment, infrastructure, amenities, design and construction. Over 80% of the participating inhabitants agreed that the main issue was poor acoustic insulation between the houses and between rooms. A more detailed study was needed to find possible solutions to these problems within a broader contextual understanding of the home.

The subsequent POE study in 2017 studied 40 homes within a section of the neighbourhood, using a questionnaire based on the inhabitants' previously identified priorities for change, home tours and monitoring of seven homes.[12] The questionnaire covered the following categories:

1. Design
2. Construction system and materials
3. Maintenance
4. Services
5. Internal layout
6. Adaptation and refurbishment

7. Adaptation for commerce
8. Comfort
9. Privacy
10. Previous housing

Results

Around 32% of households had expanded their overly small living spaces through self-build interventions, demonstrating high rates of resilience and adaptability despite having inadequate resources (see Figure 10.3). In the warm season 70% of inhabitants found their homes too hot and used some form of cooling device, ranging from fans (63%) to full air-conditioning (3%). By contrast, in the cool season, 38% of the inhabitants found their homes too cold. Temperature levels measured inside the homes during the cool season varied between 25°C and 31°C during the day, with relative humidity varying between 36% and 66%. Around 50% were still dissatisfied with the acoustics inside the home as a key issue.[13]

The average energy consumption in each home was 1410 kWh/pa (or 43 kWh/m²/pa), ranging from 419 kWh/pa to 2365kWh/pa, and 142 litres per day in water consumption. The variation depended greatly on home expansion, the number of people in the home, and activities within the home. These figures are roughly half the consumption rate per capita compared to the

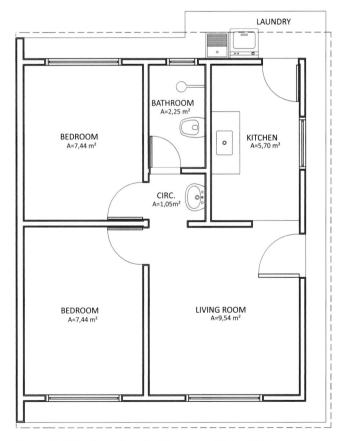

Figure 10.1 The floor plan of a Minha Casa Minha Vida home

average Uberlandia city rates, illustrating the relative poverty of this particular neighbourhood. This made the resilience of the local community all the more impressive.[14]

The recommended noise reduction for the shared wall between two homes was rated at 45 dB, but in many cases the actual reduction in acoustic tests in the homes was only around 31 dB and in one case as low as 24 dB, demonstrating the poor sound insulation qualities of the block wall. An examination of the drawings and the home tours revealed that this shared wall actually stopped at the ceiling level. This meant that the shared attic space, with its hard upper surface of the ceramic roof tiles, performed like a sound box, reflecting and transmitting noise between the two homes.

Impact

The co-production exercise used to frame the subsequent POE study as action research had three key impacts. First, the POE study had automatic buy-in from the inhabitants, who were keen to be involved in the next phase of intervention to test physical solutions for the acoustics issue identified. Second, as a UK academic, I found myself in new cultural territory working on this project and had to learn to adjust rapidly. I was able to take this learning back to my institution to help internationalise our curricula. Third, the work was carried out by researchers working with Master's students, and the co-production methodology is now firmly embedded in the school of architecture at the Federal University of Uberlandia. It can also be used more widely with other social housing communities in Brazil. As Dr Simone Villa, the leader of the research group MORA in this school, highlights:

'MORA is a physical and virtual space where several discussions about dwelling take place. It aims to be an open space for critical reflection and seeks a greater relationship between academia and practice through actions that effectively contribute to improving the quality of housing. The aims are to make information about social housing available, identifying aspects to be improved in new projects undertaken by the private sector and government with the intent to amplify the quality of the built environment in question. These experiments can promote a real, practical, difference to residents in Brazil and protect the future, providing detailed guidelines for better housing projects in a local context, proven through a study on POE.'

Figure 10.2 A typical Minha Casa Minha Vida home in the Shopping Park development

Housing Fit For Purpose: Performance, feedback and learning

Figure 10.3 About a third of the inhabitants in the study had extended their homes

International new build BPE: Australia

IMPACT OF INTERVIEWS + MONITORING

Background

One useful BPE case study from Australia questions building regulatory standards, as well as covering self-build detached houses,[15] a category often missing from housing POE studies despite forming about half of the housing stock in Australia and about 10% of the stock in the UK.

This 2009 POE study examined five Architectural Institute of Australia award-winning houses, situated individually in Darwin, Brisbane, Melbourne, Sydney and Adelaide and built between 2000 and 2003. The houses chosen deliberately reflect the variety of different climates and cultural conditions within the continent. These naturally ventilated homes used localised heating only or none at all, and all had insulated corrugated metal roofing. Cooling was by fan or natural ventilation. Each house had a different wall construction, including brick and timber frame, concrete block and timber frame, concrete block and steel sheeting, steel frame and sheeting, as well as rammed earth/concrete block and steel sheeting (see Figure 10.4). Four of the homes had photovoltaic and solar hot water systems. The sizes of the homes varied from 92 to 175 m² – much larger than the small Brazilian homes in the previous study.

The purpose was to compare the actual performance achieved against the standards in the Australian Building Code and to try and elicit the reasons for any differences.

Methods

The inhabitants and the architect were interviewed in-situ using semi-structured interviews, fully transcribed and coded. Each home was monitored hourly over a 12-month period for humidity and temperature conditions externally and in all the main living areas and bedrooms, as well as for overall energy and water consumption. There was also hourly monitoring of external wind speed and solar radiation. The results were then compared to the NatHERS simulated predictions related to the current building code.

Results

In all cases, the homes consumed much less than the average energy consumption in homes in the same locality (see Figure 10.5). At the same time they all failed to achieve the recommended indoor temperature range set out by the building code. Despite this, none of the inhabitants interviewed expressed particular discomfort, choosing to live a particular lifestyle to accommodate their sustainability aspirations.

Impact

This POE study used its findings to carefully unpick the assumptions made by the modelling software and building comfort standards adopted for Australia, and showed that people were actually quite comfortable with living in homes that were apparently 'substandard' in terms of their performance against the regulatory comfort standards. This was due to the variety of adaptive comfort measures available, including the cooling effect caused by higher airspeeds from the fans used by the inhabitants or present in the passive design of the house itself (e.g. natural ventilation and shading). The researchers recommended amending the building code and expanding the definition of comfort itself.

The interviews revealed inhabitants and architects motivated by a more holistic consideration of sustainability, including material and water resource use as well as architecture that is responsive to its place, rather than just the narrow target of energy efficiency/carbon emissions. This affected their practice of living in their house, enabling a deeply interactive and sustainable relationship to develop between the inhabitant and the building. The narrow energy and comfort criteria in the building code predicted the performance of the building within a fixed temperature zone that did not suit the inhabitants, who were used to living in both warmer and cooler conditions than specified by the standards. The researchers realised: 'The problem is that in a

Figure 10.4 Self-built detached homes form 50% of housing in Australia and are very different from UK homes

given climate these outcomes depend on both building *and* the user'.[16] They also concluded that building criteria need to take account of inhabitants' goals and architects' intentions, rather than simply fixing a rigid standard into which these aspirations must fit. This is profound, as it challenges the status quo of trying to produce simulations with generalised assumptions, which tend to generate performance gaps rather than solve them.

In some cases, the inhabitants deliberately did not insulate their home in order to be able to afford 'non-regulatory' environmental measures such as renewable energy systems or additional water storage capacity. Ironically, the building code would have forced them to specify insulation they did not need, and forgo these positive environmental measures. The findings from this study also have a wider significance, in relation to the growing adoption of these limiting regulatory energy and comfort standards by developing countries. Many of these countries have similar climates to those described in this POE case study. The research findings provide evidence highlighting the urgent need to take account of adaptive comfort measures within all these newly developing regulatory systems.

Interestingly, further research investigating comfort conditions in low-income Australian households came to similar conclusions concerning the need for a regulatory approach that takes more account of inhabitants' practices.[17] This case study also highlights the need for BPE standards to take account of local conditions and associated cultures, rather than assume that 'one size fits all'. In several of the houses examined here, the inhabitants in the tropics were used to living without the usual barriers between the outdoors and indoor living found in cooler climates – but this study recognised that inhabitants in different climates have quite different expectations, which range beyond basic comfort factors.

Comparison of actual annual energy use and annual energy use of average houses in the same region

House	Annual total energy use[a]	Annual energy use in average houses in the region with the same number of occupancy and fuel used (GJ)[b]	NatHERS predicted total heating and cooling energy load[c]	NatHERS Star Rating achieved (out of zero to 10 Stars); 5 Stars is the minimum Building Code of Australia (BCA) requirement
1	23.9 GJ (194 MJ/m²)	36.6	49.8 GJ (404.8 MJ/m²)	4.6
2	14.7 GJ (84 MJ/m²)	33.4	39.1GJ (322.6 MJ/m²)	2.4
3	23.3 GJ (141 MJ/m²)	30.8	8.3 GJ (71.4 MJ/m²)	4.5
4	4.4 GJ (27.2 MJ/m²)	26.6	26.2 GJ (186.1 MJ/m²)	4.3
5	4.4 GJ (27.2 MJ/m²)	28.3	n.a.	Zero

Figure 10.5 The homes consumed less energy but did not meet the building code

International retrofit guarantee

Background

A highly innovative development in housing is the Energiesprong retrofit programme, launched in the Netherlands in 2013 after significant consultation with inhabitants. This is the first-ever performance-guaranteed housing programme globally. It relies on a robust POE monitoring regime to prove net-zero energy use to its inhabitants after the retrofit elements are in place, to improve the image and performance of the home. This completely new approach demands real rigour and places a significant contractual onus on the combined contractor–design team to deliver the performance.

The retrofit process provides an existing terraced home with external insulation and cladding, new doors and triple-glazed windows. An external 'intelligent' service module provides the all-electric heat, ventilation and power sources alongside a new solar PV renewable energy system. A new kitchen and bathroom are also provided. The process generally takes less than 10 days to complete on site, and the inhabitants do not need to move out. If any performance gaps arising from the contract are not solved, the contractor either pays the net present value of the financial detriment for 30 years, or a pre-agreed fixed penalty (see Figure 10.6).

The net-zero guarantee means a renovated dwelling will not consume more energy for heating, hot water, lights and appliances than it produces from the installed renewable energy systems, for a period of 30 years based on a fixed service package of heating and hot water. If the inhabitants use more than this amount, then they pay for the extra usage. The €40,000 to €70,000 needed to fund each home retrofit is provided upfront by a bank loan to a housing association, and paid back through a bundled 'rental/energy plan' fee collected from the inhabitants, which costs less than the original rental fee. This innovative business model of guaranteed 'paying nothing for home energy bills' has proved a powerful incentive for the participating housing associations and inhabitants alike.[18]

Methods
The continuous post-occupancy monitoring consists of hourly readings related to energy use in the home that the inhabitant can see via a dedicated website:

- Net consumption – kWh imported energy minus kWh exported energy
- Generation – kWh energy generated from the home
- Space heating – temperature in main living room and bedroom
- Hot water – litre/kWh/minutes
- Electricity allowance – kWh used by additional lighting and power circuits
- Energy services – kWh energy consumed by running energy generation systems, heating, hot water and ventilation appliances

Results
Phase 1 of Energiesprong demonstrated savings of two-thirds of the original energy consumption by using energy efficiency measures, with the balance generated using renewable energy.[19]

Ronald Rovers makes an interesting point, however, in relation to balancing the embodied energy of a deep retrofit with renewable energy in use:
'... for a row house, we found out that the best option was to just do the cavity wall insulation and double glazing and provide the rest with solar panels. Since we did that, I am not in favour of Passivhaus style when retrofitting. I think we need to evaluate more the concept of each house to find the optimum for the materials and energy point together.'

The 'Zero on the Meter' monitoring process also stimulated inhabitants to be more economical in their energy consumption based on the following:

- The performance guarantee contract, which clarifies what is economical usage and what is not
- Feedback on usage data
- A competition element – the achievement by inhabitants of Zero on the Meter within their energy bundle agreement

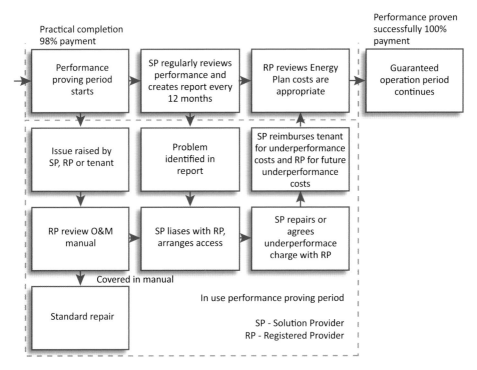

Figure 10.6 The Energiesprong process, with built-in performance evaluation

Rovers is cautious, however, about combining the predicted zero-energy performance of the house with a lump energy sum for inhabitants to use:

'I don't think you should combine the two. Because now you have set the standard for household energy use at 2,700 kWh, which means that there is no incentive to reduce that energy use or get rid of your laundry machine, or whatever. So, I would separate that, and in future, when energy prices go up, people will more easily be able to give up the laundry drier that they have because it's not included in the business deal they have for the whole house.'

Impact

Energiesprong's communication strategy takes account of the inhabitants' perspective first and does not come from technological prescriptions or the developer's assumptions. Trained 'energy coaches' encourage inhabitants towards economical energy usage. This radical retrofit approach using POE to verify performance had a significant impact on local authorities in France and the UK, where demonstration projects commenced in 2016 with a view to rolling out this approach nationally.

Mapping of techniques deployed internationally

A global review of 146 POE research projects undertaken from 2010 to 2017, and 16 associated POE protocols, identified a variety of POE methods (see Figure 10.7).[20] The use of occupant surveys was easily the most popular method, occurring in over 80% of all identified projects – almost twice as often as any other method. By contrast, only 25% of the projects undertook any form of energy calculation. This unfortunately shows how often quantitative energy studies are separate from qualitative POE studies when they should be combined for a more complete understanding of housing performance. Acoustic and water usage measurements were the least-used methods, reflecting a poor understanding of how critical these issues are. The key POE protocols are mainly from the UK, USA, Australia and Canada with relatively few identified for Asia, where studies tend to focus more on using the subjective measures only, particularly in Malaysia and Turkey. China has recently developed a POE protocol at Tsinghua University in Beijing.

The above review is relatively limited and does not include many commercial housing BPE studies that are not in research journals. Nevertheless, it provides a useful mapping of POE techniques used globally. While there is no universal agreement on the methods to be used, the following core techniques can be transferable, providing due care is taken of the context in which they are used:

- Occupant survey
- Interviews
- Walkthroughs/observations
- Energy assessment
- Water assessment
- IEQ measurements (thermal, lighting, air pollution, acoustics)

Having demonstrated the importance of BPE case studies in terms of developing methodology and challenging policy and practice, the next chapter explores how the basic techniques above and other innovative methods have been used to similar effect in several BPE case studies in the UK.

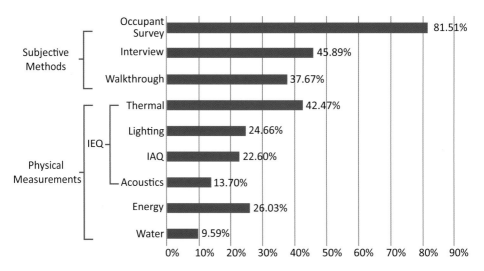

Percentage of projects that used certain method (note that most projects used more than one method).

Figure 10.7 A variety of methods are used for BPE around the world, but some are more popular than others

ELEVEN
THE UK CONTEXT

" It sometimes occurs to me that the British have more heritage than is good for them. In a country where there is so astonishingly much of everything, it is easy to look on it as a kind of inexhaustible resource. "

Bill Bryson,
Notes from a Small Island

This chapter takes a closer look at the UK in terms of BPE practice. There is a detailed focus on one retrofit BPE case study and one newbuild BPE case study that the author has been closely involved with, to provide an 'insider' perspective which is critically reflective and at the same time offers up new insights into how BPE is carried out and why.

The UK is one of the leading countries for developing housing BPE. This is perhaps not surprising, given that it has some of the poorest and oldest housing stock in Europe, an unforgiving maritime temperate climate and a high rate of fuel poverty. All of this makes it particularly challenging to maintain housing that is likely to suffer at some point from cold, mould, poor air quality or fabric decay. This has resulted in civil society being sensitive to the quality of UK housing. As a consequence, the government has funded major research and development programmes for demonstrator projects in the housing sector which contain post-occupancy evaluation (POE) studies to verify designed performance and provide insight into the performance gaps arising.[1] These programmes have provided a solid foundation over the last decade for the development of various institutional BPE policies and methods.[2] Nevertheless, there are significant contextual challenges for BPE in relation to UK housing construction, contracts and legalities that set the scene for the case studies discussed here.

The following aspects are discussed in this chapter:

- Construction, contracts and legalities
- UK retrofit BPE
- UK newbuild BPE

Construction, contracts and legalities

Housing construction
The UK has unique housing construction in terms of its existing stock and the manner in which homes are built, renovated and managed over time, all of which conditions how BPE has developed there. The Industrial Revolution in the 19th century meant that clay brick construction, supplemented by stone and timber, dominated house building in England and Wales. Terraced two-storey housing spread rapidly as a result. By contrast, Scotland had developed an earlier tradition of four- or five-storey stone tenement housing in the cities, largely inspired by continental development in Europe. After the Second World War, housing changed rapidly with the introduction of prefabricated homes for those displaced during the war, as well as the development of standardised construction using concrete elements, inspired by Modernist ideals from the 1920s and 1930s.

During the last half of the 20th century, house builders in the UK rapidly adopted timber-frame housing methods, but quickly abandoned them in England in the 1960s, after a massive technical failure to avoid interstitial condensation and rot in the construction. Scotland continued to use timber frame predominantly, while England and Wales adopted concrete block and brick construction, as well as prefabricated concrete systems for urban high-rise apartment blocks. Today there is a resurgence in structural timber for housing, based on new technology and sustainable development principles. UK housing remains exceptionally diverse in tenure, construction, form, condition and age, which presents a challenge for BPE in terms of simplifying and standardising methods (see Figure 11.1).

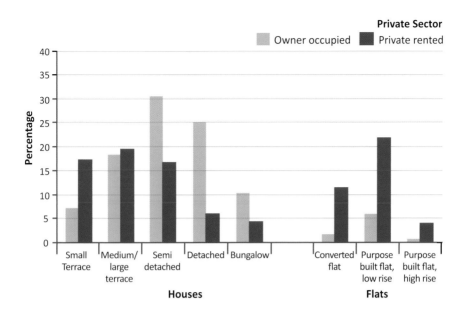

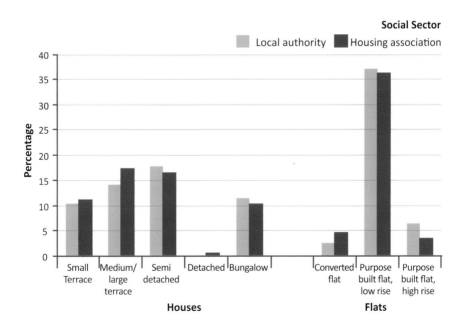

Figure 11.1 The various English housing typologies according to tenure in 2017

Contracts

Contractual arrangements in UK housing have a major impact on BPE in terms of the different protocols that affect the continuity of BPE processes. Contracts tend to fall into two types: 'traditional' and 'design and build'. The traditional contract maintains the design team intact throughout the housing project, from the initial appointment right up to the completion stage. Meanwhile the design and build contract generally removes the architect's role after the design stage – well before completion of the homes – and passes this role on to a project manager. This dislocation of the original design architect from the construction stage makes it difficult for BPE to work effectively. There is no obvious feedback route from the construction and occupancy stages back to the design architect. As a result, a crucial feedback loop is missing for improving the original housing design. Equally, the project manager has no interest in improving someone else's design work – they just manage it. Neither contract model provides specific clauses for BPE, apart from certain installation, commissioning and performance tests at the completion stage. All of these contractual arrangements therefore present real barriers for the uptake of BPE. There is, however, plenty of scope for these contracts to include the necessary protocols for BPE in practice, as evidenced by the UK Soft Landings process. Emerging 'guaranteed performance' contracts such as those developed by Energiesprong in the Netherlands and the UK also support BPE.

Legalities

Housing contracts are bound by the UK legal system. Although regulations on energy use in housing have driven up insulation and airtightness standards over the years, there has been relatively little emphasis on the actual thermal comfort provided, with no legal requirement for minimum or maximum temperatures to be achieved within a home. Equally, indoor air quality remains poorly defined, and commissioning requirements are still relatively minimal compared to those for non-domestic buildings. With this 'light-touch' approach to domestic building regulations, these issues are left to the voluntary sector to cope with via action on fuel poverty and the like. There is little scope in the current legalities of the building regulations to help push BPE forward in relation to these issues in a contractual situation.

The current regulatory system is also highly disaggregated, separating out related issues: Resistance to Moisture (Part C2), Energy (Part L), Fire (Part B), Access and Use (Part M) and Ventilation (Part F). The BPE evaluator must be aware of the conflicts that can arise between different regulations and how these can affect housing performance. Where conflicts cannot be resolved, a compromise solution may have to be recommended.[3]

UK retrofit BPE

The importance of retrofitting housing

The turnover of new housing each year in the UK is only around 1% of the existing stock. The situation is similar in many other developed countries. Housing is responsible for around 15% of all carbon emissions in the UK, and that makes retrofitting of the existing housing stock crucial to mitigate climate change. The English Housing Survey 2013 identified 9.2 million houses built between 1945 and 1980 as having an average EPC at the band D level (see Figure 11.2). The legislated requirement to reduce overall carbon emissions to net zero by 2050 will not be achieved in the housing sector without a massive and deep retrofit programme, upgrading roughly 500,000 homes a year to achieve an EPC level C as a minimum.[4] BPE plays a key role in ensuring this target is achieved in reality.

Figure 11.2 Nine million homes like this built between 1945 and 1980 need upgrading

Retrofit is different

Carrying out a BPE study on a retrofit housing project is, however, very different from a newbuild project. First, the inhabitants may be moving back into their property, which will tend to generate a different response from those moving into a newly built home.[5] Second, the monitoring process can be more complex, depending on the degree of retrofit, and the degree of consistency between dwellings in a project, in terms of size, fabric and services improvement. It can also be difficult to normalise the POE results. Third, and more positively, retrofit BPE can demonstrate actual improvements in performance, not merely hypothecated simulations. This is vital for validating retrofit strategies and tactics, particularly in relation to energy savings and CO_2 emission reductions.

A baseline BPE monitoring exercise before the retrofit can verify the improvements made and indicate what additional retrofit improvements may be needed, such as larger windows, if the lighting quality is poor.[6] A dialogue between the existing inhabitants and the design team can help to co-design the retrofit and the means of evaluating it more effectively. Through this, inhabitants can also understand what is going to happen and how to engage with the new technologies proposed. It also gives the design team insight into the inhabitants' practices and expectations,[7] allowing the retrofit and post-occupancy monitoring process to be 'calibrated' to some extent in relation to individual circumstances.[8]

Despite the benefits of carrying out pre-retrofit POE, design teams and developers tend to focus more on modelled retrofit possibilities, rather than paying attention to the existing condition of the homes they are upgrading and how the inhabitants live in them already. Pragmatic and cost-effective pre-retrofit POE surveys should be built in from the outset of any retrofit project.

Retrofit BPE case study

Background

With 4.7 million homes built before 1919 and still inhabited, the housing retrofit challenge is particularly significant for this older housing stock in the UK. The government's Retrofit for the Future demonstrator programme (2009–13) aimed to address this issue through innovative design, build and evaluation. In one of the 86 projects completed, a BPE study examined whether a deep retrofit of a typical 19th-century end-terrace house could achieve an 80% reduction in carbon emissions using the UK average figure of 97 $kgCO_2/m^2/yr$ for a 1990s semi-detached house as a baseline for comparison (see Figure 11.3). The aim was to achieve a reduction to around 17–20 $kgCO_2/m^2/yr$, with a primary energy target of 115 $kWh/m^2/yr$. A further aim was to see how effective the proposed design and process was in reality.[9]

Methods

This BPE study included a 'pre-retrofit' as well a 'post-retrofit' POE study to help understand inhabitants' practices in relation to energy use and take account of these factors in the design process. It also provided a true 'before and after' performance comparison in reality, based on the following:

- Standard Assessment Procedure (SAP) rating for energy use and carbon emissions
- Existing building condition survey
- Air permeability
- Thermal imaging
- Energy and water bills with periodic meter readings
- Levels for indoor temperature, relative humidity, lighting, carbon dioxide and thermal comfort
- Appliance energy audit
- Inhabitant questionnaire and semi-structured interviews

Additional data collected afterwards included handover evaluation, walkthroughs, inhabitant activity logs, thermal resistance and transmittance in the north walls of the home, photovoltaic (PV) and solar thermal meter readings, MVHR intake and exhaust temperatures, and window opening patterns.

Results

The pre-retrofit POE revealed that the gas consumption was half of that estimated by the SAP tool. This shows just how inaccurate the assumptions made by this particular modelling tool are, with no account taken of inhabitants' heating practices in terms of a 'pre-bound' effect.[10] The physical survey methods helped the designers to pinpoint the key areas of performance gaps to target, such as heat loss at the junctions of different materials. Thermal imaging also revealed that the external insulation to the rear was more effective than the internal insulation to the front (see Figure 11.4). CO_2 levels were high (1300 ppm), which suggested the retrofitting of an MVHR system to provide continuous fresh air. The daylighting levels were also poor, which led to a redesign to increase light levels.

Figure 11.3 The BPE retrofit study house front façade

Figure 11.4 The thermal image of the front façade showing improved insulation

The two-year post-retrofit monitoring exercise revealed that the air permeability was still too high, despite retrofit measures, to warrant an MVHR system – an expensive item costing around £6000 and using up precious energy unnecessarily.[11] The MVHR system did not perform well and CO_2 levels remained stubbornly high inside the home at over 1000 ppm for 50% of the occupied hours, most likely due to installation and/or commissioning issues. This poor air quality was also partly due to the inhabitants following guidance not to open their windows while the MVHR was on. The handover process evaluation found that the home user guidance lacked detail and was poorly coordinated with the handover process itself. There was no opportunity for the inhabitants to try out the installed technologies for themselves and ask any questions. Design teams relying on complex technologies for ventilation purposes need to provide inhabitants with feedback mechanisms on indoor air quality, to promote energy-saving practices in the home that are also healthy. This is because humans cannot easily detect high CO_2 levels.

Impact

Despite the above issues, there was a 75% reduction in carbon emissions after the retrofit, with comfort levels maintained as per design guidance, leading to a significant increase in inhabitant satisfaction. This was due to the energy-saving technologies installed, the physical retrofit, and the inhabitants' efforts to reduce energy use both before and after the retrofit through a variety of practices. This included wearing extra clothing in the winter and being able to lower space temperatures below 'normal' standards as a result. These practices were revealed through interviews, which were invaluable in disentangling physical factors from social factors influencing performance. The reduction against the 1990s average baseline adopted for carbon emissions was only 40%, however, illustrating the significant benefit of carrying out a pre-retrofit energy audit in order to obtain the real reduction figures for an individual home. This case study demonstrates the usefulness of a forensic BPE approach and the importance of using a pre-retrofit POE study in order to be able to properly understand the post-retrofit POE results.

UK newbuild BPE

Newbuild BPE case study

The political context of BPE is important to understand, and is highly contingent on the initial set-up of the study. Findings are often left out of the public story, because of the client's desire to tell only the 'good news', but sometimes this can get in the way of communicating and acting on important outcomes. This next case study explores these issues and explains how to resolve some of them.

Background

The award-winning LILAC (Low Impact Living Affordable Community) co-housing newbuild development has socio-economic and ecological credentials that mark it out as an exemplar in terms of community self-build housing.[12] Completed in 2013, it consists of 20 terraced houses and apartments, built from offsite manufactured timber-frame cassettes filled with natural straw (see Figure 11.5). The low-energy homes have 1.25 kWp PV panels, water-filled radiators heated by gas boilers, and additional hot water supplied by solar thermal systems.

The development is situated near the centre of Leeds, the largest city in Yorkshire, and the sizes of the homes are typical for affordable housing in the UK. At the time of the BPE study, around 45 inhabitants shared a variety of resources, including a common kitchen, dining area, multi-purpose room, laundry, car park, gardens, park land and allotment area.

Figure 11.5 The LILAC co-housing development

The purpose of the BPE study was to evaluate the performance of the development against the design intentions, to trial a new version of the usability tool discussed in Chapter 8,[13] and to develop a new social learning tool for housing communities.[14]

Methods

The study commenced during the handover period. This enabled the BPE team to evaluate the handover and commissioning of homes as it happened. Twelve months of monitoring followed according to a strict programme (see Figure 11.6, overpage).

The comprehensive BPE data included the following:

- SAP and EPC ratings for energy use and carbon emissions
- Existing building condition survey checked against drawings and specifications
- Thermal imaging and thermal bridging calculations (Y-values)
- Energy and water bills with periodic meter readings
- Levels for indoor temperature, relative humidity, lighting, acoustics and carbon dioxide
- Appliance energy audit
- Information on inhabitant satisfaction via a questionnaire and semi-structured interviews
- Design team intentions via semi-structured interviews
- Usability survey
- Handover evaluation and inhabitant guidance review
- Equipment installation and commissioning review
- Home visits and walkthroughs every 10 weeks

BuPESA workload plan and programme

Month 2013-15	Mar	Apr	May	Jun	Jul	Aug	Sep
1. BPE service, tools and project implementation							
1.1 Orientation and detailed planning of project							
1.2 BPE Framework and service review		M1					
1.3 POE usability and social learning tools development				M2			
2. Construction audit							
2.1 Initial meetings with the design team and client				M3			
2.2 Data capture of drawings, specifications - comparison							
2.3 Design team interviewed in relation to design intentions							
2.4 Photo survey of construction fabric							
2.5 Evening meeting with occupants							
2.6 Occupancy agreements							
3. Post-construction stage and occupancy evaluation							
3.1 Thermal imaging audit of construction fabric							
3.2 Thermal imaging and airtightness report							
3.3 Airtightness testing x 1 during thermal imaging							
3.4 Comfort and control questionnaire x all homes							
3.5 Installation and commissioning processes checked							
3.6 Home user guide evaluated							
3.7 Handover processes evaluated							
3.8 Feedback/recommendations to developer (initial)							
3.9 Portable monitoring kit ordered and installed							
3.10 Monitoring of homes for one year (energy, water, IAQ)							
3.11 Social learning tool deployed with occupants							
3.12 Walkthrough/interviews with occupants x 40							
3.13 SAP and EPC calculation review							
3.14 Y-value thermal bridging calculations							
3.15 Usability tool deployed and DomEARM audit x 1							
4. Analysis							
4.1 Semi-structured interviews/walkthrough							
4.2 Questionnaire data input + analysis							
4.3 Monitoring analysis							
4.4 Triangulation of all results							
5. Reporting							
5.1.Interim reports							
5.2 Final report (M8)							
6. Dissemination							
6.1. Internal learning for developer (presentation + guides)							
6.2 Presentations to occupants							

Figure 11.6 The LILAC BPE workload plan

ov	Dec	Jan	Feb	Mar	Apr	May	Jun	Jul	Aug	Sep	Oct	Nov	Dec	Jan	Feb	Mar	Apr	May	Jun
				M4															
		M5				M5													
																M6			
	M7												M7					M8	

Figure 11.7 There was a lack of adequate storage space in half of the LILAC homes

Results

Inhabitant satisfaction was exceptionally high compared to the BUS questionnaire average benchmarks, apart from in relation to storage provision, which half the inhabitants said was 'too little' – a general problem in new housing in the UK (see Figure 11.7).

The average primary energy usage was 100 kWh/m²/pa, and compares well with the Passivhaus standard of 120 kWh/m²/pa. However, Passivhaus is based on an average home temperature of 21°C, which not all LILAC households needed, as they preferred to wear more layers of clothing instead. The LILAC primary energy usage figure also did not take account of solar energy generated, with the PVs alone generating 950 kWh per household on average. The split of the common house energy use added significantly to the electricity usage for each household (between 10% and 25%), which needs to be weighed up against the benefit of having additional shared resources.

The main BPE team recommendation for the community was to lower its water usage, which averaged 113 litres per person per day (pppd) against a target of 105 litres pppd and a UK average of 150 litres pppd. The inhabitants were encouraged to have shorter showers for the double benefit of saving energy and water. Identifying community 'water champions' was also suggested as a collective way to share best practice.

Social learning about home technology use occurred mainly through trial and error, and random conversations. However, this led to myths generated through repetition in the community. Without the capacity to monitor and control technical systems, faults stayed unnoticed, myths went unchallenged, and some inhabitants settled into inappropriate energy and comfort habits. The BPE team recommended more structured collective learning in relation to technologies in their homes, drawing on the knowledge of the excellent community maintenance team that LILAC had established.

The most critical BPE findings, however, related to the MVHR system. MVHR air-change rates varied by up to six times from room to room within the same dwelling, and were below building regulations requirements. The MVHR commissioning looked suspicious, showing exactly the same readings recorded for each dwelling from a particular typology. The MVHR units were found to be underspecified and installed with too much flexible ducting, resulting in ducting resistance that in turn required higher fan power, leading to excessive noise. Insulation missing from the MVHR external air-supply ductwork caused condensation to drip inside the fabric of the buildings and threaten the composite straw panels with potential rot.[15] The overheating issues in the upper floors were partly because the MVHR systems could not cope effectively in the summer.[16] Eighteen of the 20 households opened their windows in the summer to overcome this issue, despite the noise, security issues and insects. One hidden and underlying reason for all of this, revealed through interviews and document reviewing, was the cost-cutting decision by LILAC to exclude the original architect and engineer from the project beyond the design stage. This meant that there was no independent oversight of the contractor's work, and it proved to be a crucial mistake.

Impact

Challenging the contractor and the contractor's M&E engineers over the various MVHR issues identified needed careful handling. The situation was delicate, with LILAC building up its reputation as a new community and keen to establish its credentials. In order to move forward, the BPE team instigated a process of 'double accounting', with issues reported immediately for internal actioning by LILAC with back-up support from the team, rather than being publicised any further. In the end, the contractor agreed to recommission all 20 MVHR systems, as well as taking down all the kitchen ceilings and installing extensive amounts of missing insulation on the MVHR ductwork. Good negotiating skills were essential for the BPE team, who needed to address LILAC's lack of knowledge, while at the same time encouraging and supporting them to challenge the contractor.

During the study, some LILAC members requested raw energy data from the BPE team, prior to the BPE team having analysed and published their own findings, which could have invalidated the team's unique research. The resolution was to provide LILAC with extensive analysed data to help the community understand what was happening in their homes, while protecting the intellectual property of the BPE team. Subsequent regulations under the UK General Data Protection Regulations (GDPR) have clarified the right of researchers to retain anonymised personal data where the research itself may otherwise be jeopardised. Chapter 13 discusses the ethics of such data use in more detail.

The delicate results of the BPE study were not publicised by LILAC themselves. Instead they were publicised by the researchers through conferences, research papers and book chapters. BPE was not a particular priority for LILAC at that time, given that it had many other self-imposed environmental and social targets to achieve within a relatively stretched resource base.[17] The BPE study had been proposed by the researchers rather than LILAC. Nevertheless, LILAC made significant changes based on the BPE findings and their members also learned how to operate their homes more effectively during the course of the study, as a form of action research.[18]

The two case studies in this chapter illustrate the adroit use of skills and methods in BPE to address unexpected results, in terms of inhabitants' and contractors' expectations and practices. The retrofit case study also shows a pre-design BPE process to improve proposed retrofit interventions. The newbuild case study is an example of how BPE can help clients and inhabitants to learn how to optimise the performance of their homes, as well as holding a contractor to account, despite the lack of statutory regulation relating to actual performance. The next chapter examines another critical factor in achieving successful BPE – its cost.

TWELVE
THE COSTS AND BENEFITS OF FEEDBACK

" What is a cynic? A man who knows the price of everything, and the value of nothing. "

Oscar Wilde

Is BPE really worth it? A more useful question might be 'what is the value of BPE?' This enables a broader discussion of the intangible benefits that BPE can have in terms of creating new understanding to aid housing improvements. This extends beyond individual project recommendations, to help to de-risk and futureproof the national housing stock as a whole. In this way, BPE can add considerable value to the housing design and build process as well as the occupancy and maintenance of homes. Nevertheless, cost generally remains a major barrier to BPE in the UK. Government programme BPE requirements, for example, have proved overly expensive for house builders and their design teams.[1] BPE is seen as too expensive because the cost has to be built into the capital programme and the benefits are not realised to the house builder directly through revenue.

This perception needs to be re-balanced in terms of capital cost and revenue gains. The case for BPE can usually be won when a sound argument for good business value is presented to the client. The case should be made well ahead of their housing project in order to build in funding as part of the budget. BPE can also be presented as a bonus opportunity during or after the build stage, providing alternative funding is available. Most house builders are unwilling to undertake BPE studies once their development projects are completed, without these measures, as their budgets are already committed.

Based on my personal experience, this chapter provides a critical cost analysis of three different housing building performance evaluation/post-occupancy evaluation (BPE/POE) projects to help identify best value at different scales. It also looks at the wider, less tangible value of BPE. The following aspects are covered in this chapter:

- Who pays and who benefits?
- Costing feedback into practice
- Costed examples
- Valuing BPE

Who pays and who benefits?

National government, agencies and local authorities currently pay for BPE studies via various national funding programmes and local grants. They typically benefit in terms of increased GDP and reduced subsidies needed for housing, as well as reduced health and service costs due to action taken in relation to BPE findings.

House builders pay for BPE studies by match-funding government programmes, or by obtaining funding from government agencies or other sources and adding in their own time and resources. They benefit in terms of improved reputation, products and processes, and reduced risk and liability.

Design teams often pay indirectly for BPE studies, in a similar way. They reap similar benefits to the house builders, as well as learning how to improve their design practice through feedback. Practices are also setting up their own BPE consultancies as a paid service.[2]

Academics and consultants primarily benefit from BPE through research, and are paid directly to carry out studies. They also benefit from learning continuously through their own BPE practice.

Inhabitants rarely pay for BPE directly (though they contribute, through taxation, to government-funded grants), unless they are very intrepid homeowners. However, they too give their time for free to help the studies. They reap significant economic, social and environmental benefits, described later on.

Costing feedback into practice

The three primary funding routes for housing BPE are as follows:

- Design practice
- Housing developers
- Academic partnerships

Practice-based research and development

Professional institutions are now actively encouraging their members to engage with practice-based POE in order to improve their organisational knowledge and reputation,[3] drawing on a variety of funding sources. Practices can also deduct their staff costs for research against corporation tax.[4] This provides a useful means of initiating and integrating a Soft Landings approach at a relatively low cost. Various architectural practices are already using an informal approach to BPE at minimal cost.[5] However, rigorous BPE as a routine activity requires practices either to absorb the cost within their continuing professional development (CPD) budget, or build it into the costs they pass on to their client. Practices also need to invest in BPE training and embedded organisational learning to make BPE part of a virtuous circle of design improvement.

The cost of the comprehensive LILAC BPE study (£57,700) was just less than 2% of the overall housing project cost of around £3m in total in 2013.[6] For larger developments, the costs reduce further due to efficiency gains. An architectural practice typically paid around 5% as a design fee for housing will find the cost ratio of this BPE study very significant. In most cases, however, a 'light-touch' BPE study costing between £5000 and £10,000 will be perfectly adequate to start with. A practice can initially undertake BPE only on its larger housing commissions, before making BPE routine on every project, in order to maintain profitability.

Housing developer-initiated BPE

Housing developers who take advantage of government grants in order to undertake research and development to improve their products typically 'match' a government grant on a 50/50 basis, with 'income in kind' activities. This can be a very effective way to subsidise BPE processes in housing development. When external funding was not available, some developers indicated in 2013 that they would expect to pay no more than around £5000 to £10,000 for BPE study. This suggests that evaluators should probably adopt a light-touch rather than an investigative approach in the first instance. Another means to embed BPE costs routinely is via the Energiesprong retrofit model described in Chapter 11.

Academic partnership

A typical route for a practice or housing developer to engage with BPE is via an academic partnership. Funding generally comes from research or knowledge transfer/exchange grants provided by the government or charities, and can cover most of the cost of the BPE work. Take care to cost in the time needed for the design team and the client, as these grants often only cover the researcher's costs, with the design team and client having to contribute in kind. Sometimes PhD or Master's students can carry out a BPE study, under supervision of an experienced academic, free of charge as part of a 'live project' or dissertation partnership, with the cost of the BPE kept within the cost of the student fee for the taught programme. Most of these one-off partnerships tend not to lead to BPE embedding in practice, however, as there is usually no research follow-on funding to enable this. However, one way the embedding can occur is over a series of discrete research projects carried out consecutively in partnership.

Scale, skills and overheads

Regardless of the funding route used for housing BPE, there are issues of scale, skill levels and overheads to consider when budgeting for BPE. How many homes to sample within a single housing development for BPE purposes, to be both rigorous and cost-effective, depends on whether the purpose is to:

- pre-test a prototype home in order to design out the performance gap
- improve performance at every stage of a housing project on site; or
- verify the post-occupancy performance of the finished product.

For initial prototyping, only one home needs to be evaluated. For performance improvement, one selected sample for each typology needs to be built ahead and evaluated to provide a site 'role model' for the rest during the construction stage. For verification of performance, a certain percentage of homes (typically around 10%) should be evaluated, and ideally all inhabitants in the development should partake in a satisfaction questionnaire. For benchmarking and other statistical purposes a minimum of at least 25 homes need to be monitored. The bigger the sample, the greater the level of certainty in terms of results (see Figure 12.1).

In a study of 39 domestic BPE projects in the Technology Strategy Board (TSB) UK BPE programme (2010–14) the average total cost per BPE study was £61,978 for an average of 3.5 homes evaluated at a forensic level, which included £19,142 for monitoring equipment supplied and installed.[7] Individual study costs varied significantly, depending on the type of housing, equipment, methods and overheads.

Balancing the skills required for the various BPE tasks, and the associated overheads, can make a significant difference to the cost of a BPE study. For a basic study on a small number of homes, a single experienced BPE evaluator can offer excellent value for money. The evaluator should be independent from the design team and not overly influenced by

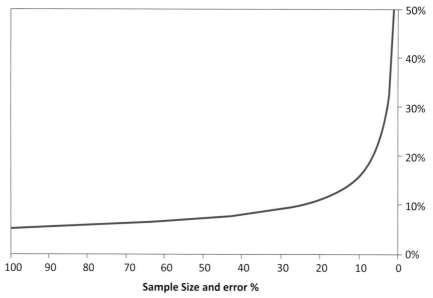

Sample Size and error %

Figure 12.1 A sample of 25 homes for monitoring will give a good indicator of certainty for any large housing development

their desire to gain good results. A small BPE team with relevant design, technological and social survey expertise is best for studies that are more complex. Within this team, junior assistants can help set up equipment, collect data and analyse it, under the supervision of a more experienced evaluator, as another means of reducing costs.

One often-forgotten overhead is the depreciation of equipment. It is quite usual to 'write off' monitoring equipment at the end of a two-year monitoring project, given how rapid the depreciation is and how quickly the technology is changing. Hiring equipment can be one way around this. Cost savings on equipment occur with BPE across a number of homes rather than a single home.

The next section discusses indicative budget costs for POE in three selected case studies from three different funding sources. Personnel costs and equipment costs are highlighted separately.

Costed examples

Light-touch: Kincardine O'Neil
The first phase of this POE study of 12 homes (see Chapter 6) was carried out on a budget of just £4,500 in 2002 using a small grant from a regional housing agency in Scotland. This bought 18 days of the single consultant BPE evaluator, with an additional £500 for travel and consumables, and no equipment costs. It was a 'light-touch' project using seven POE methods (see Table 12.1). The local pub kindly donated its dining room as the 'site office' from which all field work was conducted. A simple interviewing technique used voice recording and note taking, not the full transcripts as usually required. When combined with a review of the maintenance log this gave rich data for the minimum cost. In hindsight, the note taking may have biased the interviews. This was mitigated to some extent by listening to the voice recordings while checking the notes.

Table 12.1 Comparison of BPE personnel costs for three projects

BPE methods (including analysis time)	Light-touch: Kincardine O'Neil (12 homes)	Diagnostic: LILAC (20 homes)	Forensic: Sigma Home (1 home)	BPE activity hours per home
Design intentions				
Design team 30-minute interview	√	√	√	7
Document review	√	√	√	7–63
SAP check		√		4
Fabric survey				
Construction audit	√	√	√	7–70
Thermal bridging calculations (Y-values)		√		21

BPE methods (including analysis time)	Light-touch: Kincardine O'Neil (12 homes)	Diagnostic: LILAC (20 homes)	Forensic: Sigma Home (1 home)	BPE activity hours per home
Thermal imaging		√	√	7
Co-heating test			√	49 (Sigma Home) – 140 (LMU)
U-value check			√	49
Home tours/ walkthroughs	√	√	√	3
Moisture in fabric measurement				
Commissioning				
MVHR system check		√	√	4
Heating system check			√	4
Renewable energy systems check		√	√	4 per system
Air permeability test checks			√	14
Handover				
Handover evaluation		√	√	7
Guidance evaluation		√	√	7
Monitoring over 12 months				Varies according to scale and intensity
Weather station			√	7 to install
Temperature		√	√	
Humidity		√	√	
CO$_2$		√	√	
VOCs			√	
Energy – metered		√	√	
Energy – bills	√			
Water – metered		√	√	
Water – bills				
Window/door openings			√	

BPE methods (including analysis time)	Light-touch: Kincardine O'Neil (12 homes)	Diagnostic: LILAC (20 homes)	Forensic: Sigma Home (1 home)	BPE activity hours per home
Infra-red sensing for occupancy factors			√	
Occupancy				
Acoustic tests – spot only			√	14
Lighting tests – spot only			√	4
DomEARM energy audit		√	√	7
Questionnaire (e.g. BUS)*		√*		100 questionnaires* 28
Interview inhabitants 30–60 minutes	√	√	√	7–12 per interview
Focus group inhabitants				9
Social learning		√		Varies
Video of usability			√	4
Usability survey		√		14
Maintenance logs	√	√		7
Inhabitant diaries/logs			√	14
Thermal comfort survey			√	Varies
Ethnographic observation		√	√	Varies
Reporting				
Meetings with client	√	√	√	14–28
BPE report (+ interim)	√		√	35–70
Dissemination				
Inhabitants' meeting		√	√	7
Publicity	√	√	√	Varies
38 methods total	7 methods	23 methods	30 methods	

BPE methods (including analysis time)	Light-touch: Kincardine O'Neil (12 homes)	Diagnostic: LILAC (20 homes)	Forensic: Sigma Home (1 home)	BPE activity hours per home
Total hours cost	£4,500 (2002)	£53,200** (2013)	£21,000 (2008)	
Normalised personnel cost 2018 (UK Consumer Price Index)	£6491	£58,137	£27,011	

* 20 homes surveyed for inhabitant questionnaire
** Extracted pro rata from overall research costs

There was no attempt to prove the fabric performance of the homes, which would have been relatively costly. The benefit of the rich data obtained from a minimal number of key methods was significant, however. Numerous emergent issues surfaced, including undersized mechanical ventilation units causing mould, lack of acoustic insulation between floors, and uncontrollable 'total control' night storage heaters – the energy supplier controlled these rather than the tenants. Although not all recommendations were actioned, the lessons were conclusive for the client and design team. The energy costs proved inconclusive, however, as the bills covered different periods, which could not be resolved. Monitoring of the meters would clearly have been more effective, but also more costly and intrusive at that time.

Diagnostic: LILAC

In this POE study of 20 homes (see Chapter 11) the £53,200 personnel costs were funded by a relatively large European Union research mobility grant in 2013. This allowed for a much more comprehensive selection of methods (see Table 12.1) that were based on the the TSB Building Performance Evaluation programme protocol for POE (Technology Guidance Version 7). The co-heating test requirement was omitted as it was too disruptive for the inhabitants. Several innovative POE methods were added as part of the research project: usability evaluation, social learning evaluation (see Chapter 8) and maintenance log evaluation.

Several intangible benefits here are difficult to account for in terms of cost. In this action research project, the inhabitants were learning about how to use their homes as the findings emerged. Equally, the researchers were learning about how inhabitants in a relatively new housing typology (co-housing) were living in their homes. In hindsight, the budgeted costs also considerably underestimated the time spent on liaising between the inhabitants and the contractor to address the performance issues discovered. It is important to factor in these extra costs for BPE projects that involve intensive engagement with the inhabitants and clients.

Forensic: Sigma Home

The prototype Sigma Home project described in Chapter 8 included an intensely forensic BPE study, commissioned by the house builder (Stewart Milne Group) to inform the future design of a completely new range of housing typologies.[8] The house builder obtained Energy Saving Trust funding for the POE activities undertaken by a small team of two researchers, budgeting for £21,000 personnel costs, as well as £550 travel/consumables costs (see Table 12.1). However, the client probably put in at least as much time in terms of responding to the action research with troubleshooting, which should be borne in mind for this particular type of BPE project.

The personnel costs are considerably more expensive for evaluating this one-off house compared to the TSB BPE programme average (£12,239 per home, of which about 24% was for administration, project management and reporting to TSB). However, this type of project helped solve design issues for a completely new product range for the house builder that added significant value. In hindsight, the BPE time was probably undercosted by a factor of two, due to the sheer amount of time and effort needed to deal with the complexity of this innovative home.

Equipment costs

Costing BPE equipment is an art as well as a science, given the rapid development of technologies and the constantly changing protocols as a result. It takes an experienced BPE evaluator to judge what the best selection is for any particular project. Many BPE projects have ended up with unforeseen equipment costs. This is often due to adopting a generic approach for all BPE projects, only to discover that different forms of construction require different types of equipment, and different sizes of housing development need different equipment approaches for aggregating data.

BPE protocols, such as the Energy Saving Trust's (EST) 'Monitoring of New Homes'[9] or the TSB's BPE programme protocol, can predetermine the expected type and performance of equipment for POE studies. The EST approach suggests that basic monitoring equipment would only cost about £1500 per home (at 2008 prices) for temperature/humidity sensors (three internally and two externally) and a data logger, using pre-installed smart metering for water and energy monitoring. The assumption is that the sensors and datalogger are already wired in during the build stage for any housing development. Monitoring of renewable energy systems costs extra per home, for example photovoltaic panels (£300), solar thermal panels (£650) and a heat pump (£550). By comparison, the average ongoing monitoring equipment cost per home in the TSB BPE programme was £5469 (priced between 2010 and 2013), based on a much more forensic approach.[10] However, costs are coming down.

There were no equipment costs for the Kincardine O'Neil study. In fact, there is much to commend not using additional equipment, beyond that already installed, for an initial light-touch POE study of a housing development or home. This is because the equipment itself can become the problem in the first instance, and distract the evaluator from attending to the overall preliminary evaluation of the performance of the physical home in relation to how it is being lived in. It is only necessary to 'drill down' with the deployment of appropriate equipment if any significant issues are revealed. If the inhabitants are happy and the home is performing as expected, then further testing and ongoing monitoring are redundant.

The monitoring equipment costs for the LILAC study were minimised by using cheap 'I-button' sensor/data loggers for measuring temperature only in the living room, and Hobos sensor/data loggers for measuring temperature and humidity in one bedroom and bathroom, in all 20 homes. A CO_2 sensor/data logger was installed in five homes. An existing local weather station provided external climate data. Manual readings were taken from the existing water, gas and electric meters every 10 weeks while the data was being downloaded from the sensors, removing the need to install any additional meters. The MVHR testing equipment and thermal imaging camera were hired from a local university. Thus, the total equipment cost reduced to £4500 (£500 per house).

The three quotes obtained in 2007 for the monitoring and testing equipment used in the Sigma Home varied wildly, from £4790 to £7885 for the two wireless systems, and £10,500 for a standard wired system. In the end, choosing the cheapest system proved to be a mistake. This was because it was a beta version of a state-of-the art wireless monitoring system that did not deliver as expected and required back-up (at additional cost) with traditional monitoring equipment for half of the monitoring time. Testing innovative monitoring equipment in BPE studies is probably unwise unless the team and the client are prepared to meet any extra costs arising from this.

Valuing BPE

So just how valuable is BPE? Cost-benefit analysis (CBA) in BPE means identifying the financial, social and environmental costs and benefits of carrying out different types of BPE studies compared to not undertaking one. Each benefit relates to the people, housing, organisations and environments immediately involved, and also other people, housing, organisations and environments where these benefits can have an impact. This gets extremely complicated and perhaps explains why there is no obvious CBA carried out for BPE approaches. Interestingly, some academics are now beginning to undertake and critique CBA in relation to sustainable housing using POE methods.[11] The costs of BPE are considered next in terms of some of the benefits that the three studies above generated, rather than undertaking a full-blown CBA.

Capital versus revenue costs: the business case

A good business case for housing BPE can be made by comparing the cost of a BPE study against the future revenue savings for a housing developer, contractor and design team in terms of reduced defects, maintenance, risk, liability and futureproofing.

For example, the LILAC BPE study revealed that the inhabitants and client had assumed that their solar panels did not need cleaning in urban environments, even though they were not steeply inclined. There were no roof access or ladder anchor points for the solar thermal and photovoltaic panels, and there was no way to tell whether the solar thermal panels were malfunctioning, potentially causing thousands of pounds' worth of damage.[12][13] LILAC had to pay for a 'cherry picker' crane and operative (around £500 a day) to enable safe access to the roof, every time a problem occurred or the panels needed cleaning (see Figure 12.2). The discovery of condensation from the uninsulated MVHR ductwork dripping on to the straw-based structure also prevented further thousands of pounds' worth of future damage, as well as significant future disruption for the inhabitants, and effectively de-risked this technology for the client.

The BPE study of the Sigma Home helped the house builder to significantly reduce the cost of its new range of homes by highlighting the excessive complexity of the technologies deployed and the redundancy of having 21 opening windows, when only five were used in practice.[14] When this factor is multiplied over thousands of planned units, the cost saving is significant compared to the cost of the study. A recommendation to avoid placing solar panels on the walls, where they would potentially be disabled by future homes overshadowing them, again saved thousands of pounds through futureproofing the typology (see Figure 12.3).

Physical benefits

BPE studies can improve fabric and services performance by highlighting design, specification and process issues overlooked by the design team and contractor. For example, thermal imaging of the Sigma Home revealed an inferior glazing panel installed in the windows. In another BPE study for the Lancaster Cohousing group, we discovered that sawdust produced by the contractor's carpenters had lodged in the MVHR system via its air intake, which had been deliberately left switched on during the final stages of the contract to help dry out the plasterwork in the homes (see Figure 12.4). This significantly reduced the efficiency of the ventilation system, but was easy to fix once diagnosed. The BPE study also revealed a large amount of missing insulation at the roof edges, which was quickly installed once diagnosed, preventing possible condensation issues through cold-bridging.

Figure 12.2 The LILAC BPE study revealed that easy access to the PV panels on the roofs had been removed at the design stage as a cost saving

Social benefits

Improved liveability, health, wellbeing, comfort and empowerment occurs when BPE studies verify and promote best practice as well as identifying issues that need to be addressed. Some social benefits are difficult to cost (e.g. inhabitant empowerment through learning how to use technologies), but important issues related to health can be calculated in relation to poor indoor air quality, lack of warmth and mould in homes. Improving this has significant cost benefits for the National Health Service and employers. Using BPE to retrofit appropriate energy saving measures (new boiler, double glazing and insulation) reduced GP appointments by 60% and Accident and Emergency attendances by 30% in one UK study.[15] Social return on investment (SROI) based on traditional cost-benefit analysis is a relatively new method for BPE. It assigns monetary values to social benefits and compares them to the level of investment. SROI used in recent POE studies has still to capture all the social impacts, however.[16]

Environmental benefits

Reductions in energy use, carbon emissions, pollution and waste/resource use due to BPE help to increase the sustainability of homes as well as mitigating climate change. Calculating the primary in-use and embodied carbon emissions saved after BPE interventions is often a good proxy for these environmental benefits. In a study of low-carbon housing in Rotherham, the BPE team discovered that one home was using 19,000 kWh of electricity a year, with an annual bill of £2500. Diligent work diagnosed a poorly commissioned and faulty heat pump. A rapid BPE intervention resulted in a 30% reduction in energy use the following year.[17] As part of a comprehensive BPE project to evaluate the use of unfired eco clay bricks in a prototype house in Scotland, I analysed the embodied energy of these bricks. At 25.1 kg/CO_2 per tonne the eco bricks saved 7 tonnes of CO_2 compared to using common fired bricks for a three-bedroom house.[18] This BPE study proved the case for using these bricks in future housing design and specification, and led to their subsequent manufacture in the UK.

Figure 12.3 Vertical solar wall panels can quickly become obscured by neighbouring activity

Figure 12.4 Carpenters unwittingly released sawdust into these MVHR air inlets, blocking the filters

Valuing the intangibles

Reputation is a key driver for developers and designers. It underlies profit margins. As an intangible asset it is hard to gain and very easy to lose. This has been demonstrated by UK housing disasters in the past, such as the partial collapse of a high-rise apartment block in London, killing two and injuring 17 in 1968, and the mass interstitial rotting of timber fabric in homes in England in the 1970s. More recently, the Grenfell Tower fire disaster led to the death of 72 people and injured 70 in 2017. These faults arise from poor design, construction and inspection regimes that can be improved by integrating housing BPE processes. Conversely, housing quality is another intangible asset that BPE helps to promote. It is very difficult to cost these two intangibles, but the risk of losing them make the case for BPE that anticipates future problems even more compelling and valuable.

BPE also has a valuable multiplier effect through dissemination of its findings beyond the initial client to policymakers and subsequently into the industry. A good example of this was the Stamford Brook BPE project undertaken by a team from University College London and Leeds Metropolitan University, which revealed that the lack of insulation in the party wall between terraced or semi-detached homes was producing a 'thermal bypass' (see Figure 12.5), which significantly reduced predicted performance[19]. This finding changed the national building regulations and housing practice generally to ensure that all party walls are now fully insulated. BPE knowledge transfer reduces industry-wide risks, and helps to lower the premiums for professional indemnity and project insurance, as well as reducing the risk of litigation by alerting the industry to practices that are not working as intended.

Linking BPE findings to Housing Quality Indicators, such as those in the BRE Home Quality Mark, increases their value by driving evidence-based design quality across the industry. At present neither POE nor BPE is a mandatory requirement in the Home Quality Mark, which is unfortunate and potentially unethical, considering the housing disasters described above. At the highest level, BPE can provide a way to help evidence and develop progress towards the UN Sustainable Development Goals as means to a prosperous society and as a moral imperative. This leads us into the consideration of ethics in BPE practice, as discussed in the next chapter.

Before   **After**

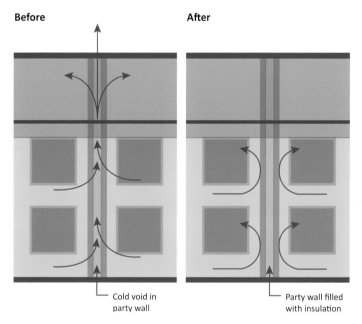

Cold void in party wall

Party wall filled with insulation

Figure 12.5 A thermal bypass process in party walls was discovered through BPE

SECTION

5

CHALLENGES FOR FUTURE

THIRTEEN
THE ETHICS OF FEEDBACK

" Ethics determine choices and actions, and suggest difficult priorities. **"**

John Berger

Ethics represent a system of moral values and principles that affect how people make decisions and lead their lives. They cover rights and responsibilities in terms of what is good and bad for individuals and society. There are a number of ethical issues involved in BPE, which need to be resolved before undertaking a study.

Housing BPE is unavoidably intrusive to people's privacy, as it can reveal individual behaviours, habits and lifestyles. The intimacy of a home is much more personal than a non-domestic setting. Inhabitants take the evaluators into their trust, and it is vital that this is fully respected within the duty of care provided for any study. The ethics in this bond are complex. The BPE evaluator aims to verify design intentions, which the inhabitants may not have been consulted on. At the same time, the study should involve the inhabitants, while maintaining their anonymity, to reveal useful insights. These dual aims can lead to delicate situations in which the vested, and sometimes contradictory, interests of all parties involved must to be taken into consideration when BPE work is undertaken and reported.

Handling the ethics of BPE is challenging because new issues are constantly emerging in conjunction with the development of new technologies and new environmental, social or economic policies. Seemingly 'neutral' monitoring of energy data can actually reveal personal behaviour in great detail. Drone cameras can be even more intrusive. How should such monitoring be policed?

This chapter explores various dilemmas present in the ethical management of housing BPE, and discusses how these can be resolved within the housing BPE approaches and methods described in Chapter 7. The following aspects are discussed in relation to the process of carrying out a BPE study:

- Ethical BPE evaluation: do no harm and be honest
- Ownership of data
- Recruitment, consent and participation
- Sharing data
- Benchmarking and metadata
- Framing the feedback

Ethical BPE evaluation: do no harm and be honest

Having a robust ethical policy as part of a BPE study helps the team to navigate ethical issues successfully. The policy documentation needs to include seven principles as a minimum:

1. No purposeful harm
2. Honesty and integrity
3. No coercion
4. Informed consent, including a right to withdraw
5. A requirement for confidentiality
6. Equality and diversity
7. Data protection[1]

Universities in the UK have robust ethics procedures defining engagement with research participants that provide excellent guidance for use with BPE and are available on university websites. There is, however, an art to minimising the amount of text in an ethics policy to ensure that it is acted on. When confronted with 30 pages of university ethics application forms, housing developers and design teams can feel overwhelmed. The UK Research Integrity Office (UKRIO) therefore has a number of useful resources related to ethics, including a simple one-page checklist that can be used at the start of designing a BPE study (see overpage).

A checklist for ethical issues to cover in BPE

1. Does the proposed research address pertinent question(s) and is it designed either to add to existing knowledge about the subject in question or to develop methods for research into it?

2. Is your research design appropriate for the question(s) being asked?

3. Will you have access to all necessary skills and resources to conduct the research?

4. Have you conducted a risk assessment to determine:

   a. whether there are any ethical issues and whether ethics review is required;

   b. the potential for risks to the organisation, the research, or the health, safety and wellbeing of researchers and research participants; and

   c. what legal requirements govern the research?

5. Will your research comply with all legal and ethical requirements and other applicable guidelines, including those from other organisations and/or countries if relevant?

6. Will your research comply with all requirements of legislation and good practice relating to health and safety?

7. Has your research undergone any necessary ethics review (see 4(a) above), especially if it involves animals, human participants, human material or personal data?

8. Will your research comply with any monitoring and audit requirements?

9. Are you in compliance with any contracts and financial guidelines relating to the project?

10. Have you reached an agreement relating to intellectual property, publication and authorship?

11. Have you reached an agreement relating to collaborative working, if applicable?

12. Have you agreed the roles of researchers and responsibilities for management and supervision?

13. Have all conflicts of interest relating to your research been identified, declared and addressed?

14. Are you aware of the guidance from all applicable organisations on misconduct in research?

When conducting your research:

1. Are you following the agreed research design for the project?

2. Have any changes to the agreed research design been reviewed and approved if applicable?

3. Are you following best practice for the collection, storage and management of data?

4. Are agreed roles and responsibilities for management and supervision being fulfilled?

5. Is your research complying with any monitoring and audit requirements?

When finishing your research:

1. Will your research and its findings be reported accurately, honestly and within a reasonable time frame?

2. Will all contributions to the research be acknowledged?

3. Are agreements relating to intellectual property, publication and authorship being complied with?

4. Will research data be retained in a secure and accessible form and for the required duration?

5. Will your research comply with all legal, ethical and contractual requirements?

Source: http://ukrio.org/publications/

Our personal values and consequent actions are bound up with societal values involving trade-offs as a way of trying to strike a balance between conflicting goals and desires.[2] Currently, certain societal values are increasing the negative effects of climate change on humans and other forms of life. The BPE evaluator is thus in the double bind of trying to behave ethically within an inherently unethical situation. Professional institutions in the built environment need to give higher priority to mandating wider environmental and social responsibilities in their codes of conduct. The RIBA's Ethics and Sustainable Development Commission Final Report in 2018 recommended that POE should be mandatory for RIBA-registered practices within two years, and that associated research methods should be taught on RIBA-validated programmes.

Within a BPE project unintentional harm may be done either to the home itself (for example the disruption of the construction fabric to install monitoring equipment – see Figure 13.1) or to the inhabitants (e.g. through the revelation of 'anonymised' personal details which may be recognised by others from a small housing study). Evaluators need to make every effort to avoid these risks, and make sure that everyone is aware of all the known risks involved in a BPE study. Ultimately, a collective judgement call must evaluate the overall risk and benefit of any action. Sometimes honesty and integrity can be unavoidably compromised (e.g. undiagnosed monitoring equipment faults leading to false feedback). The unwillingness of a housing client to publish findings may also compromise their reputation, with the collusion of the BPE evaluator in the perpetuation of poor practice through this lack of dissemination.

A positive approach to ethical professionalism is set out in 10 key principles that BPE evaluators can follow as described by Bill Bordass and Adrian Leaman[3] (see below).

Elements of a new professionalism

1. Be a steward of the community, its resources and the planet. Take a broad view.

2. Do the right thing, beyond your obligation to whoever pays your fee.

3. Develop trusting relationships, with open and honest collaboration.

4. Bridge between design, project implementation and use. Concentrate on the outcomes.

5. Don't walk away. Provide follow-through and aftercare.

6. Evaluate and reflect upon the performance in use of your work. Feed back the findings.

7. Learn from your actions and admit your mistakes. Share your understanding openly.

8. Bring together practice, industry, education, research and policymaking.

9. Challenge assumptions and standards. Be honest about what you don't know.

10. Understand contexts and constraints. Create lasting value. Keep options open for the future.

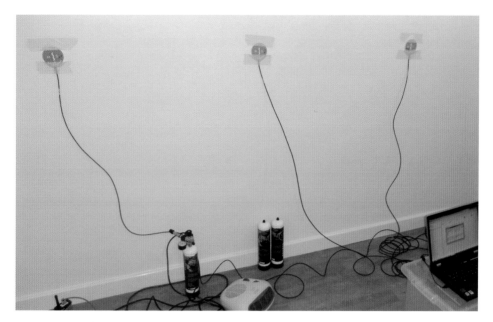

Figure 13.1 Monitoring equipment can disrupt wall fabric – an ethical issue

It is also essential for BPE evaluators to keep their professional knowledge and skills up to date and to give stakeholders the information they need in a format that they can understand easily.

Ownership of data

The first step in developing an ethical BPE study is to define ownership of the BPE data and any restrictions relating to this. This must be recorded in the BPE contract documentation and include what will be collected, and how it will be collected, stored and used, as well as who will have access to it and what their responsibilities are in relation to the data.

BPE includes physical, organisational and personal data. Evaluators must decide a) what data is strictly necessary for the study and b) what data needs to be kept beyond the study for wider purposes.

Physical data

This relates to the housing development, homes, materials, assemblies and products. Typically it consists of physical testing and monitoring of energy, water, indoor air quality, building fabric, climate information, specifications, samples, photographs, drawings and other artefacts. This data is often the easiest to define contractually in terms of ownership. Sometimes physical data can become personal data; for example, if a thermographic image or photograph of someone's home can be identifiable, it could be deemed to be a personal identifier. Equally, the house-builder may not wish to be identified with this image. Care needs to be taken to ensure physical data is suitably anonymised where required (see Figure 13.2).

Organisational data

This relates to the housing developer, the design and build team, funding agencies, suppliers and installers. It is important to clarify the ownership of any organisational data provided by the various parties to the study (such as information about design methods, management policies, organisational systems, websites, etc.) as well as who owns the intellectual property rights (IPR) to the study. This must be set out in the BPE contract, and requires negotiation between the parties. This may involve apportioning ownership of IPR, or granting a licence to use the outputs in perpetuity while ownership remains with the client.

Personal data

This data relates to all those involved in the study. The General Data Protection Regulation (GDPR) (EU) 2016/679 law, approved by the EU in 2016 and adopted by the UK, sets out the requirements for how personal data is to be used. 'Personal data' means any physical, physiological, genetic, mental, economic, cultural or social identification of a specific natural person. BPE evaluators need to prove that there is a legitimate interest in carrying out data collection and in keeping any personal data confidential from those offering it up. The presumption is that personal data is not withheld from the person involved. This includes items such as their individual electricity bills, physical monitoring records, interview notes and transcripts.

 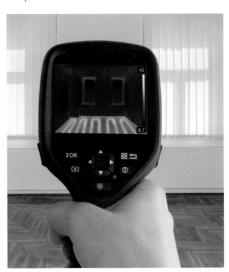

Figure 13.2 Thermographic images: one identifiable and one more anonymous – the difference is important ethically

Recruitment, consent and participation

Recruitment

Recruiting inhabitants to a BPE study by providing proportionate payments or small tokens for participation, depending on their level of involvement, is ethical. They should not be coerced and must have the right to withdraw at any stage. It is a good idea to recruit inhabitants via the house builder, housing landlord or a community organisation, as they can usefully broker any questions or issues related to the study, particularly if they are already known to the inhabitants. Another effective means of recruitment is via a community meeting held near the proposed BPE study, where the study is explained and a question and answer session is offered to address any concerns. This can also help with more effective co-design of the study, as inhabitants often come up with useful contributions.

Informed consent

Obtaining informed consent is essential in BPE at all times. A simple one- or two-page information sheet and a consent form must be provided to each participant. These should clarify in plain English (and other languages if necessary) what the BPE study is about, what the participant has agreed to in terms of the collection, use and storage of their data, what information will be shared with them, and whether the study will lead to any action on issues identified. Send the forms to the participants well ahead of any planned BPE activity to ensure buy-in. They can be delivered digitally or by post (with a prepaid return envelope) or through a third party (e.g. the housing landlord). Alternatively, circulate them immediately before an activity commences, providing the participant has previously indicated that they intend to sign the consent form. The consent form must be signed and retrieved before any activity commences. It must then be safely stored for the duration of the project and any associated activity beyond it.

Participation

The level of participation will depend on how intensive the BPE study is, and the degree to which participants are encouraged to co-design the study. The action-research nature of the LILAC BPE study meant that there was significant participation by inhabitants both on an individual basis and collectively.[4] The researchers visited individuals in their homes every ten weeks over a year, as well as meeting the whole community several times to discuss the set-up and findings of the study as it developed. Take care to manage expectations in terms of problem-solving, when there is intimacy and trust between the BPE team and individual inhabitants. All dates for home visits should be agreed well in advance, and a risk assessment must be in place to protect the health and safety of the team as well as the inhabitants.

Sharing data

Two key questions for data-sharing agreements are 'Who is the study for?' and 'Who is funding it?' If it is the same client, the BPE team should negotiate the agreement with them directly. A BPE study funded separately by a government agency, research organisation or charity, which supports the client commissioning the work, will have separate agreements relating to organisational data sharing, reporting and dissemination. In collaborative BPE studies involving several parties, all have to agree to the terms of sharing data. In all cases, the right to privacy and personal data for the inhabitants must be respected, although there are times when higher considerations may override this, as discussed next.

Duty of care issues

There will always be 'duty of care' situations, where a team must report a critical health or safety issue to the client, and share data. This includes observing practices that are seriously damaging the fabric of a home and threatening the health of the inhabitants, such as drying significant amounts of washing indoors while blocking off all ventilation equipment and keeping windows shut, leading to excessive mould developing on the walls. It is essential to report these issues to the client, as there may well be other reasons why they are occurring, due to faulty design, installation or commissioning issues such as noisy ventilation systems, or lack of insulation in the walls, which the client then needs to investigate. The BPE team may be placed in a difficult position unless the boundaries of such reporting are agreed from the outset.

Sharing data and reporting

Data from BPE may also need to be shared between organisations providing different BPE skills, and it is essential that this is covered by contractual documentation that clarifies what data can be shared between these parties, when and how.

A BPE reporting strategy should establish who will produce what type of report and when, what the approval stages are, and what degree of confidentiality is required for each stage. This part of the contract documentation will determine whether the reporting is to be internal, and whether publications can follow on to disseminate the work.

Sharing data publicly, including monitoring data, makes it possible to develop a more robust BPE knowledge base using large open-source datasets for validation. This requires a careful procedure to ensure that the shared data is compatible with other existing datasets of the same kind, and that the type of licence for usage, including any embargoes on publication, are clearly set out.

The degree to which personal data is anonymised is another important consideration. There are various levels of anonymisation used when data is processed. The BPE team need to communicate to participants the level of anonymity being used. This applies particularly to questionnaires, which can appear to be anonymous in terms of findings arising from them, but which may contain personal identifiers to aid processing. This is so that the researchers can identify who returned which questionnaire. The consent form should cover the use of anonymised interview quotations as well as anonymised photographs and video clips from inside and outside homes to avoid revealing the identity of the inhabitants and their homes. The evaluator must obtain specific consent for use of images and video clips that cannot be anonymised (e.g. showing the façade of a home to highlight fabric issues).

Publications

Sharing the hard evidence from BPE studies more widely means that housing sectors around the world can learn how to prevent the spread of poor practice, such as the faulty installation of thousands of MVHR units in recent housing developments in the Netherlands and the UK.[56] Getting approval to publish BPE findings, however, requires tact and good timing. Many house builders ask for a 30-day approval period for any proposed publications, so that they can vet them. Inhabitants need to be consulted on these publications if they are not anonymised. It is possible to write a publication that anonymises the findings while still communicating vital lessons, to the extent that the client and inhabitants do not need to be asked for publication permission. The use of social media for disseminating BPE studies is increasing. Social media sites can be a great way for inhabitants to actively engage with BPE projects,[7] but care needs to be taken here also to anonymise personal data.

Benchmarking and metadata

Benchmarking BPE performance is an ethical issue in itself. The current use of energy efficiency as a benchmark is too narrow and does not take account of overall energy demand in relation to people's practices and the services provided to them.[8] A more ethical benchmark would be the *total* amount of energy that a home or a person actually uses as a measure of *sufficiency*, rather than just how *efficient* the energy use is per square metre. This would move BPE beyond seeing energy efficiency as the same services for less energy input, to directly tackle the unsustainable level of servicing of the modern home itself.

A gradual re-evaluation of international sustainable building benchmarking systems over time has introduced a more nuanced and at the same time broader consideration of the qualitative factors that need to be incorporated.[9] BPE teams always need to question what they are measuring and what this is being compared to. This is to ensure that they are not excluding vital needs such as aesthetics, human development and biodiversity, which are often at the periphery of BPE benchmarking categories but very evident in the UN Sustainable Development Goals (SDGs) and Human Development Index. At present, there are no robust national benchmarks for housing BPE due to a lack of consensus on performance measurement requirements. This needs urgent resolution before housing BPE can be properly legislated for.

Benchmarking exercises can also unfortunately encourage data 'cleaning', such as the removal of outliers or spikes in BPE monitoring data[10] which, when associated with qualitative data such as inhabitants' interviews or diaries, usefully explains why these deviations are occurring. Carefully selected benchmarking yardsticks help provide data appropriately controlled for context. Data should always be normalised at the last possible moment, to keep the rich context for analysis. If there is a need to control for context (e.g. climate variation), the process should be clearly explained.[11]

Once uniform performance parameters have been agreed (e.g. total KWh/a kWh/m²/ pa, kWh/person/a or CO_2e/person/a) it is also possible to generate useful metadata from aggregated shared datasets that can inform useful meta-studies on particular aspects of housing BPE. Recent studies of performance in homes in the UK have used such metadata in relation to ventilation[12] and the overall performance of retrofit projects.[13]

Framing the feedback

A discussion of the ethics of feedback would not be complete without considering how BPE feedback is framed and disseminated.

If BPE is to help the housing design and build process meet core needs, the feedback needs to go back to the design team, contractor and house builders in the first instance. Feeback to inhabitants helps them understand how to live in their homes more sustainably. Overall, the feedback should deal with the in-use footprint of the home. But the household carbon footprint may well be vastly in excess of the home-use carbon footprint, once food and travel are taken into account.[14]

How inclusive should BPE feedback be, then? Shouldn't it include aspects of travel (location of the housing development in relation to public transport connections), food (availability of a garden for growing vegetables) and lifestyle (keeping pet snakes in a vivarium may use more energy than is used for heating the home itself)? And shouldn't BPE include the full life-cycle analysis (LCA) of the embodied carbon and energy involved in the construction, maintenance and future demolition/reuse of a home as others have suggested?[15] Ethically, the answer is yes, but pragmatically, the answer is probably no – these should instead form adjacent studies to keep the BPE project itself manageable in terms of time, cost and publication. Any LCA studies may in any case be more effective when undertaken at the design stage.

There is then the question of how the framing of the feedback itself is value-laden in terms of those doing the reporting. In many BPE reports, inhabitants have been devalued and blamed for the failings of a home due to their own behaviour, when they are simply trying to cope with poor design, production, construction, installation, commissioning, handover or guidance issues. It can be easier for clients, professionals and builders to blame the inhabitants in relation to difficult BPE findings than to be more self-critical and examine the causes within their own organisations.

BPE feedback should not apportion blame to individuals (who may come and go within organisations) but rather try to pinpoint which organisational systems are failing, where the break points are occurring, and why. Apportioning blame to an individual allows an organisation to pass off a systemic or process failure as 'human error', when the system or process has clearly not been designed to be able to cope with human error, which is an inevitable activity.

When organisations have the integrity and determination to reflect on their own processes in relation to BPE findings, the results can be transformational, as was the case with Architype Architects, who improved their practice significantly,[16] and the Stewart Milne Group, who completely redesigned their housing production process.[17] At its best, BPE helps organisations to identify and continually improve the thinking underpinning their processes as well as the outcomes for all involved. This very much depends on making sure that BPE feedback loops are well set up in the first place – an area which is examined next.

FOURTEEN
EFFECTIVE FEEDBACK LOOPS

"Designers seldom get feedback and only notice problems when asked to investigate a failure."

Alastair Blyth,
CRISP Commission 2000–2

Good housing evolves by ensuring that vital feedback from design in use informs the next iteration of design and delivery. Feedback helps built environment practices thrive through the development of refined solutions, honed by informed experience. It is also in the interest of professionals to know which aspects of their practice are effective in reality, and which are not. The medical profession has always had very strong links between practice, education and research, which ensures that all three areas are based on feedback from performance. At present, this is relatively tenuous in architectural practice, which urgently needs to develop a similar culture of demonstrating performance in reality.[1]

But what makes feedback effective? How can the feedback loop be closed to ensure continuity and change at the same time?

The art and science of creating effective feedback loops within BPE practice depends on an agile combination of tacit knowledge, good negotiation skills and the ability to recognise when there is a need to call on the help of others. This is because knowledge is only ever partial and is always changing. BPE requires a continuing variety of perspectives to help build a more inclusive and accurate understanding of housing performance. In this way, best practice in design and delivery can be more effectively developed and celebrated directly through BPE.

Effective feedback, however, also depends on a willingness to acknowledge when there is a need for improvement and to be able to act on this knowledge in practice without fear of reprisal, rather than perpetuate a dodgy detail or a faulty construction process. This approach can develop into a 'no-blame' systematic learning process that ensures feedback embedded in the practice is fed forward right at the start of any project, rather than at the end of a project, when it is often too late.

This chapter sets out what is needed at each stage of a project to embed BPE methods within a co-designed feedback and learning process to enhance the practice of both the client and the design team. The following aspects are covered:

- Building feedback into the project process
- Set-up, communication and contracts
- Reporting to enable change
- Dissemination for impact

Building feedback into the project process

A housing BPE framework

The useful six-step cyclical model developed by Wolfgang Preiser and Ulrich Schramm provides BPE with a strategic framework and systemic process for the complete life cycle of a housing project and into the next one.[2] The BPE briefing requirements within a housing project work plan relate here to each RIBA Plan of Work stage (see Figure 14.1). Using this framework is vital when developing the overall housing project brief, within which the BPE approach, methods and techniques should sit as part of the BPE sub-brief. Doing this enables robust learning and development to take place related to a continuous cyclical project process in practice. The typical 'client' in a housing project, however, is the developer rather the

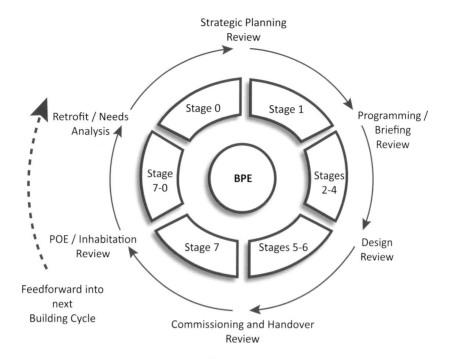

Strategic Planning
Review

Retrofit / Needs
Analysis

Programming /
Briefing
Review

Stage 0

Stage 1

Stage
7-0

BPE

Stages
2-4

POE / Inhabitation
Review

Stage 7

Stages 5-6

Design
Review

Feedforward into
next
Building Cycle

Commissioning and Handover
Review

Figure 14.1 RIBA Plan of Work stages and six-step cyclical BPE model

inhabitants. This can lead to false assumptions in the project brief about inhabitant practices that need to be fundamentally questioned from the outset, drawing on BPE feedback from previous projects, and also explored via the BPE project itself.

The **strategic planning stage (RIBA stage 0)** should include the key aims, priorities and overall requirements for the BPE work. Use a financial review to establish what is feasible within existing constraints and whether additional resources are needed to carry out the BPE process. Review the latest developments in BPE in case these prove useful. Discussion between client representatives, the design team and existing inhabitants should highlight whether or not previous BPE recommendations have been actioned, and how this will affect the planned project. If a previous BPE study is not available, an initial discussion of the effectiveness of previous housing project delivery within the client organisation and representative inhabitants is useful to undertake. Any existing housing strategies should be strategically reviewed at this stage as the first action in the BPE brief; this includes spatial, construction, health and safety, environmental and 'smart' controls, maintenance and sustainability strategies. This is essential to identify any clashes between the various strategies in relation to the planned housing project.

The **programming/briefing stage (RIBA stage 1)** is when the detailed housing BPE brief is developed, incorporating all relevant criteria, benchmarks, methods, techniques, personnel and specific costs. This should include a 'light-touch' post-occupancy evaluation (POE) approach, initially, with some contingency in case deeper levels of investigation are needed. The brief can be prepared by the housing developer working with the architect and the BPE specialist, and, ideally, with inhabitant input, or as a multi-agency brief. The performance specifications for the BPE project are now defined in the brief. This is the point to ensure that any overall design lessons from the previous BPE studies are firmly lodged in the briefing documentation. BPE continuity needs to be built in at this stage to work between all the stages. A 'BPE champion', particularly one specialising in organisational learning, will be invaluable to assist with the continuity of performance evaluation, and they should be specified in the BPE contract. Include a client signing-off point for each stage of the BPE process, in order to ensure that the outcomes are reviewed sequentially. This develops an incremental improvement process rather than delivering a bundle of unexpected 'surprises' at the end of a project, when identified problems are most difficult to rectify.

The **design review stage (RIBA stages 2–4)** can incorporate early testing opportunities into a BPE brief, such as testing prototype design solutions using physical mock-ups or even whole-house mock-ups, which can be temporarily inhabited (see Chapter 8 and Figure 14.2). Previous BPE feedback can really help here, to evaluate proposed design solutions – incorporate a review of this into the BPE brief. Eliciting the views of any existing or potential inhabitants in relation to the proposed designs provides another form of BPE 'sense-checking' and co-design opportunities.

The **commissioning and handover stage (RIBA stages 5–6)** is the part of the BPE brief that describes how to evaluate the performance of various systems and check on commissioning-related activities as part of the briefing, design, build and maintenance stages in a housing project, comparing performance against design intentions. The brief should establish the BPE evaluation criteria and methods used to check the commissioning

Figure 14.2 The design review BPE stage can be used to test housing using physical mock-ups

of systems, if needed, and to demonstrate compliance of all service systems, in relation to any commissioning and maintenance plans. The brief should also cover criteria and methods for evaluating the project handover and inhabitant induction processes and outcomes. Commissioning and handover lessons from previous BPE studies should be built into the brief.

The **POE/inhabitation stage (RIBA stage 7)** sets out in the BPE brief the testing and evaluation of assumptions made in the previous stages. The BPE project should also evaluate and facilitate inhabitant understanding and optimisation of the use of their home. Organisations should aim to include ongoing cyclical light-touch BPE reviews as part of their housing project aims, priorities and overall requirements, to enable continuous improvement of the outcomes related to feedback. These reviews should happen in tandem with scheduled maintenance events, major refurbishment points, and re-selling of the property, to ease the burden on the inhabitants.

The **retrofit/adaptation stage (RIBA stage 7, linked to 0)** should include in its BPE brief an evaluation of all design options for retrofit, adaptation or deconstruction for reuse and recycling, drawing on lessons from previous BPE studies (see Figure 14.3). The BPE process should evaluate how well this stage has worked in relation to the original intentions.

Two additional resources that can also help with strategically planning the BPE brief are the RIBA's own guidance on Soft Landings and POE, and guidance provided by the Usablebuildings website (http://www.usablebuildings.co.uk/).

Figure 14.3 Retrofit BPE should evaluate options for deconstruction to reuse materials and elements

Set-up, communication and contracts

Set-up

The first move towards setting up a housing BPE project usually comes from either an in-house or independent BPE team, or a client. On rarer occasions, a third party brokers the potential partnership – for example, a government agency providing a workshop where potential clients and BPE/design teams pair up. The difference is important, as it sets the tone for subsequent negotiations and the form of the final BPE contract agreed. In all cases, the client will need to be convinced of the BPE team's credibility. This is best conveyed in a meeting between the parties via a half-hour presentation on BPE and plenty of discussion time after this that can be the first step towards co-designing the BPE brief. The presentation itself should be in plain English and jargon-free, with copious illustrations to evidence previous BPE findings of relevance. A good rule of thumb is to present no more than one image a minute on average. This presentation and subsequent discussion can build on the initial written proposal from the BPE team to the client.

The art of closing a BPE project agreement lies in negotiating services and costs that have a realistic budget depending on the type of BPE work to be undertaken (see Chapter 12). Ideally, the light-touch BPE study should automatically form part of the initial design commission and an ongoing practice process, but there will be times where an additional, standalone BPE study is requested during or after a design project has been completed. In both cases, clear financial boundaries need to be drawn between the BPE project requirements and the ongoing and separate snagging, commissioning exercises that form part of an overall housing project. Without this, activities can get muddled, leading to recriminations later on, should the BPE team decide that inherent re-commissioning work is outside the scope of their activities. All the ethical issues discussed in the previous chapter need to be addressed in the scope of works and contract agreement.

Contract

Typically, a light-touch housing BPE project will be included in the architect's appointment and the client's generic contract conditions, followed by an appendix which lists the services to be provided by the BPE team, including the milestones and timescales. For additional BPE work at a diagnostic or forensic level, it is likely that an additional contract will be needed. If a BPE team has already obtained funding to do such work (e.g. via a research grant), they will tend to operate under a similar type of contract, but this time according to conditions set out by the funding organisation for the BPE team. Where there is a consortium of multiple parties an initial Memorandum of Understanding may need to be signed by all parties to initiate the confidential negotiations, followed by a signed contract.

For a light-touch BPE project, a single payment on submission of a final report will suffice. For more extensive BPE projects, there should be key milestones and associated deliverables for the different activities aligned to a payment schedule. A simple Gantt work plan is very useful to include in contract documentation. It needs to show the duration of each BPE activity, the number of hours/days involved, and when it will take place (see Figure 11.6 in Chapter 11).

Factor a contingency element into the BPE project, in terms of time and costs, to allow for any unforeseen events such as equipment failure, lack of access to homes or data loss. Insurance and indemnity will be needed to cover the BPE team and inhabitants in case of any related accidents, damage or theft during the BPE activities.

Communication

Ideally, there should be a single point of contact between the client and the BPE team to avoid any confusion in communications. It is important to keep both the client and the inhabitants informed about planned BPE activities, as well as to report back in good time to both parties on findings as they arise. For a light-touch and smaller BPE project, it is probably enough to have just one formal introductory meeting with all the relevant representatives, followed by a final meeting. For longer and larger projects it is more usual to have bi-monthly or quarterly formal progress meetings each year. Regular contact in between these meetings is useful to address any matters arising as the project progresses. This generates two-way feedback, with the client and the inhabitants helping the design and BPE team to understand the context of any particular challenges as they arise during the project, and deal with them more appropriately. One such successful dialogue occurred during a retrofit project of 150 homes in Kerkrade-West in the Netherlands, as described by Ronald Rovers. It started with '… a structured engagement with the inhabitants, involving them early in the process and convincing them of the advantage. During the preparations for the project, an employee was dedicated full time to the job for a year, organizing meetings, making house visits and involving a local committee in the whole process – and with success.'[3]

In larger housing organisations, it is worth trying to make sure that the CEO or managing director knows about the BPE project via the client representatives, and if possible to invite them to the presentation and discussion of the findings. This can make a strategic impact, and begin to expand and embed the BPE process within the organisation itself. It can also encourage repeat BPE contracts with the same organisation.

Reporting to enable change

Reporting stages
Reports and presentations are a key part of ensuring the success of any housing BPE project; too long, and the client will lose interest – too short, and important context may be left out. The timing of reports is also critical. Any interim and final reports should be timed to ensure the findings are put in front of a more influential panel, beyond the immediate representatives acting on behalf of the client organisation. This panel could be the board of directors or an appropriate work group or committee that has strategic power to act on the BPE findings in terms of developing responsive actions across the organisation.

Report content
For a light-touch BPE project write a short final report for the client, design team and contractor, extending to no more than about 20 pages with appendices as necessary. This can be written by the BPE evaluator, in conjunction with the design team. For a more in-depth project, an interim report and final report will be needed, which need to be written by the expert BPE team. The interim report should signpost any significant trends for the client to act on if necessary. There should always be a clear one-page executive summary, which briefly explains the project and captures the key lessons and recommendations. The main body of any BPE report needs to cover the following:

- Executive summary
- Introduction and details of the housing project
- Aims and objectives of the BPE project
- BPE methods used
- Basic results
- Analysed and evaluated findings
- Key lessons and recommendations related to improving the housing process
- Remedial actions

Always celebrate positive results first (see Figure 14.4). The focus should then be on unusual and exceptional outcomes rather than expected results, with a rich explanation of why these have occurred, what to do about them in terms of remedial action, and whether further investigation is required. It is often relatively easy to fix the small things that inhabitants find most annoying, such as missing draught excluders or poorly balanced ventilation systems. Lessons and recommendations need to be carefully organised in terms of priority, and in relation to the original design intentions – questioning these where necessary. Avoid pages and pages of recommendations, as these are less likely to be read comprehensively. A simple table, which places priority actions and the investment needed, in relation to the immediate (now), medium term (one to two years) and long term (five years plus), can be a very effective mobilising tool for the housing project team and client to help improve their practice. Any political sensitivity should be taken into account, with a separate report for public consumption produced if necessary.

Attach any additional sub-reports on particular BPE activities as appendices to the final overall report. Typically, these can be broken down into physical testing, monitoring and inhabitant-related sub-reports. This flexibility enables the client and BPE team to choose which sub-reports to send to different entities as required.[4]

Figure 14.4 Celebrate positive results first in BPE – Lancaster Cohousing performed exceptionally well

All the BPE reports should support findings with suitably distilled and appropriate data, maps, graphs, diagrams, tables, photographs and sketches, ideally covering about one-third or half of the pages with these, leaving about two-thirds or half as text. This makes the coverage more engaging. It is also very compelling to include telling quotations from open-ended questionnaires (such as BUS), interviews or focus groups.

Benchmarking and data sharing
In the BPE report, benchmark the results against the project design targets, other sustainable design standards and housing studies using existing databases. Where none exist, compare results with similar projects. Ideally, the BPE criteria should tie in with the

Housing Fit For Purpose: Performance, feedback and learning

organisation's overall key performance indicators and benchmarking criteria. This may be relatively easy for energy use, carbon dioxide emissions and indoor environmental quality, as well as for proprietary inhabitant questionnaires such as BUS, but can be more difficult in relation to qualitative data. Comparing context-rich quotations from residents via interviews or focus groups on key aspects of performance can greatly aid organisational learning, particularly when these draw attention to specific housing processes and design details. Sharing any personal data within or beyond the participating organisations needs to be done with due care (see Chapter 13). The benefits of this benchmarking exercise are potentially huge. It can generate more robust national BPE datasets for housing and improve the performance of homes, particularly when related to case studies and key lessons. It can also influence design guidelines for future housing projects.[5]

Dissemination for impact

Dissemination strategy

After reporting on a housing BPE project, the next step is to increase the impact of the findings. The BPE team should develop a dissemination strategy as part of their initial proposal. Ideally this should aim to embed the findings within housing, design and construction practice, as well as within built environment professional education curricula. There is a need for organisational transparency here. If there are problems and shortcomings revealed by the BPE findings, as well as successes, these should be acknowledged. Some discussion will be needed around how to live with the findings, or work around them where they cannot be immediately addressed, and how to present them in the public domain.

Make sure that the inhabitants and other housing staff involved in the BPE project are given an opportunity to see the results and findings, and to comment on these, prior to any wider publication. After all, they have contributed their time voluntarily to the project. This is best done through a presentation and question-and-answer session on site, followed up by the client giving the inhabitants access to the original reports lodged on the organisation website or, where matters are too sensitive, via a summary report gauged towards inhabitant understanding. The client representatives and BPE team should collaboratively develop the targeted presentation for maximum benefit. It can also include recommendations for the inhabitants to help them to engage more effectively with their homes and home technologies.

Embedding lessons in practice

Within the design office and housing organisation, a very simple electronic template can record key lessons referencing the BPE final report, after an initial presentation of the findings to the practitioners. Make sure this completed template, as well as the BPE final report, is accessible for everyone as part of the quality management system. This can provide an excellent form of CPD. The designated BPE champion should complete the template to ensure continuity of learning. The templates can be tagged in the office library in relation to key products and details, particularly where performance issues and solutions have been highlighted. The completed template can also link into the Building Information Modelling (BIM) process as a suitable 'object' for reference within the facilities management section of the project, and linked to the briefing documentation. To reinforce the BPE lessons, it can be useful to develop a reflective office quality-management 'wiki' system, which captures the experience of subsequent projects and attempts to respond to them. The BPE lessons will need reinforcing with regular BPE workshops, reflective team reviews and site visits to share and embed the learning experientially with the whole project team as further CPD. BPE findings can also inform further practice-based research and development of particular housing typologies, processes and products.

Wider practice impact can be generated from BPE findings through office-sponsored or public seminars and workshops, which can also help to build up a caucus of housing BPE practitioners across different organisations. Generating policy impact is best done through professional bodies and government agencies. Local politicians are always interested in housing performance, which directly affects their electorate, and are often willing to donate their time to a BPE event. There are also cross-party parliamentary groups of MPs who aim to improve the quality and performance of the built environment and who are interested in BPE.

Traditionally, the professional indemnity insurance industry has actively prevented the sharing of negative BPE information. The prevention clauses are often embedded in the insurance contract itself. This has created a culture for architects and others not to share. This is changing now, with the RIBA confirming that indemnity insurance can cover BPE. The insurance industry actually has a strong influence on the governance of BPE in the wider sense of the term. It determines the degree to which professional mistakes are allowed to be publicised or not. This publicity is needed in order that everyone can learn the lessons from these mistakes and not repeat them. It is important to encourage this industry to use its influence to support BPE more openly and prevent the sheer magnitude of misery caused by known underperforming design strategies which are not disclosed.

Wider impact

Beyond this, publicising the findings of housing BPE more widely as guidance enables learning across the housing industry as well as improving the reputation of the organisations involved as market leaders and specialists in the field of housing BPE. This involves sharing the findings at CPD events, professional seminars, national and international conferences as well as through practice pamphlets, book chapters, books, and research journal papers co-authored with academics. Offering BPE projects as case studies for built environment academics and students can also help to foster a strong relationship with educational institutions, possibly with future collaboration involving students helping out on 'live' BPE projects.

Using social media and other forms of news media is another excellent way to disseminate housing BPE findings and gain further feedback. Inhabitants and others can act directly via social media that may even move BPE as an activity from the professional building sector to other stakeholders. If architects and engineers do not actively claim BPE as a prerequisite to their work, they may well lose this domain altogether. To draw attention to BPE reports and findings, hashtags such as #buildingperformance or #housingperformance can be useful on Twitter, and referencing specific organisations in the tweet can also help. Care should be taken when dealing with the news media, however, as reporters tend to slant BPE findings towards their particular agenda. It is always advisable to ask to see a copy of the proposed text to publicise the BPE findings, and to obtain permission from the client before agreeing to this type of publicity.

Having effectively embedded housing performance feedback into practice as routine, two questions remain for this book: what is on the horizon for BPE and housing, and where next?

FIFTEEN
NEXT STEPS

" … the fact that our developers managed to get their performance gap down, showed that by focusing on quality and focusing on improving performance, you could actually address this problem … "

Jon Bootland, former
Good Homes Alliance CEO

Despite the Jon Bootland quotation, BPE is still not considered important in housing. In a telling judgement of its perceived impact, the latest UK voluntary housing performance standard, the Home Quality Mark, gives post-occupancy evaluation (POE) just 10 out of 500 possible credits overall, as an *optional* activity. So why is housing BPE not routine? What would make it more relevant and engaging? This last chapter summarises the main barriers and opportunities, and provides key lessons to help make feedback more effective. A forward-looking agenda is presented for policymakers and practitioners to make housing BPE an integral part of the design, build and management process.

The following aspects are covered in this chapter:

- Overcoming barriers
- Making feedback more effective
- Closing the loop
- Setting the agenda: 10 questions
- A final word on housing BPE

Overcoming barriers

Four key groups present particular barriers and opportunities for BPE: manufacturers, house builders, professionals and policymakers.

Manufacturers

Manufacturers hold the key to future BPE, by ensuring successful product use in housing. House builders already feed back performance issues to their supply chain, but manufacturers still rely on testing products in laboratories or using individual 'field trials' where the physical assessment of their product works in isolation from other crucial contextual factors. A richer BPE approach is needed, beyond life-cycle analysis (LCA), which tests building products with people in their homes and reveals how products and components work in tandem with other considerations. There is now a real opportunity for manufacturers, and their representative institutions, to champion a systematic BPE approach for assessing their products as they actually work in reality.

House builders

One challenging aspect of delivering housing BPE relates to the different tenures developed by house builders (including self-build), related to the following:

- Private ownership
- Private rental
- Social rental

Private ownership

Once their homes are completed and sold, there is little incentive for house builders to find out from the inhabitants how well these are performing beyond basic customer satisfaction levels. As Julia Green, a project manager and former Head of Sustainability in the major UK house builder Crest Nicholson plc, points out:

'While we go back and survey them, historically we haven't asked any questions about the performance of their home or what it was like to live in. It was more about the service they'd received, and perhaps sometimes design and layout aspects, but not what it was like to live in.'

There is immense pressure to provide a return on capital investment and move on to the next housing project as quickly as possible, which often prevents adequate reflection on the effectiveness of their construction procedures and policies:

'With an organisation like Crest, and probably every other house builder, they are at full pelt, day to day, week to week, month to month, building, building, building, building. And there almost isn't the capacity to say – "if there was a study running alongside this, it might be all very nice, it might tell us stuff" – but in a lot of ways I think it would fall on deaf ears, because the business is in complete operational mode.'

A surprisingly good time for developing BPE is when the housing market slows down and house builders have more time to reflect on their processes and invest for the future. The largest housing organisations have the capacity to take a lead and act as BPE role models for others. Adopting a win-win 'performance guarantee' approach using POE is another way to overcome current inertia (see Chapter 10).

Private rental

Private landlords are relatively disinterested in understanding the performance of their homes unless they receive complaints. Policies concerning performance therefore tend to be reactive rather than proactive, and there is virtually no BPE in this sector. A mandatory benchmarked rating system for BPE assessment and disclosure on the basis of a 'design to perform' approach would help here, similar to the NABERS initiative for non-domestic properties in Australia, where buildings are routinely expected to perform within 10% of the predicted energy use.[1]

Social rental

Social housing and build-to-rent organisations invest over the long term to provide their inhabitants with affordable, comfortable and healthy homes. They want to understand how their housing performs in order to improve it for their future inhabitants. Despite this, only the most committed organisations such as the Joseph Rowntree Housing Trust undertake BPE. Government housing agencies have supported POE activities in the past, and should always have a major role in mainstreaming BPE. National organisations representing the social housing sector need to show more leadership in this area.

Self-build

There is a major opportunity to develop a BPE approach specifically for self-builders, as it is in their own interest to ensure that their homes perform as planned. Community self-build housing groups have a collective capacity to undertake BPE studies within this sector, through co-housing groups for example. Encouragingly, individual Passivhaus self-builders are increasingly monitoring the performance of their own homes.

Professional barriers

While some professional institutions in the built environment have made tentative forays into developing BPE resources, there has been no cross-institutional collaboration to promote this activity. Various cross-disciplinary organisations have tried to overcome this issue (e.g. Green Building Councils and Construction Industry Councils) but these have relatively little influence on the institutions themselves. As housing performance and practice is intimately related to institutionalised knowledge and explicit rules, there is a significant opportunity for these institutions and others to collaborate and ensure that BPE is developed using a truly interdisciplinary approach[2] (see Figure 15.1). It is important that both insurance and financial bodies are included here as part of the 'client' base – ultimately, their policies and rules govern the housing process, including any BPE.

Professional institutions need to overcome the perceived barrier of the potential liability attached to any defective performance identified through BPE. The RIBA has already confirmed that architects with an RIBA Insurance Agency professional indemnity insurance policy are covered to undertake BPE services.[3] They could do more to help all professionals and housing organisations understand that an early identification of any defect is much better than this remaining hidden – for everyone. Apportioning BPE payment remains another unresolved barrier. Professional institutions requiring their members to undertake BPE on their built projects, particularly as this activity now forms part of the new British Standards on building information modelling (BIM) and Soft Landings, would resolve the payment issue, as it would automatically be included in the fee structure.

Policy barriers

The UK government's Clean Growth Strategy states: 'Innovation can help overcome non-financial barriers, in particular behavioural barriers, to energy efficiency. Research Councils are planning to invest around £19 million to research *how people's energy choices can help them stop wasting as much energy*'[4] (author's italics). This illustrates a classic policy 'blind spot', of continually placing the responsibility for poorly performing homes firmly with the inhabitants rather than the house builders, product manufacturers, designers, installers or building contractors. Inhabitants' choices are strictly limited by the housing products they are given, and this needs to be recognised. Industry cannot simply innovate its way out of poor performance, without first properly evaluating and learning from current performance.

Figure 15.1 The key actors involved in BPE must collaborate to make BPE truly interdisciplinary

BPE evaluators need to challenge conventional policy emphasis on changing inhabitants' behaviour rather than evaluating and improving current industry practice and technology performance. Even the design purposes themselves need challenging at times, in order to build new relationships between design, housing performance, everyday life and policymaking.[5] Policymakers should feed back energy evaluations and interventions directly into decision making and policymaking and make this routine.[6]

Making feedback more effective

Four areas offer opportunities to make BPE more effective: health, participation, prototyping and education.

Health

The identification of poor external air quality in the UK and other countries has generated renewed interest in health issues related to housing performance. There is increasing awareness of the significant affect that indoor air pollution in the home has on people's health and its link with energy efficiency. The public often perceive health issues to be more important than either environmental or housing issues, providing a quicker way forward for BPE (see Figure 15.2). As Jon Bootland, from the Sustainable Development Foundation, puts it:

'You can imagine that health and money is such a sensitive issue for so many people, so that instead of having to wait for legislation to change, to protect society as a whole, it could easily become a demand from purchasers. This could drive behaviour much quicker for developers than the other way.'

The 19th century saw rapid housing policy reforms based primarily on health grounds (Chapter 1). There is good reason to believe that a focus on health issues could kickstart

Figure 15.2 Focusing on health issues is a key driver for ensuring BPE is carried out routinely

housing BPE, including summer overheating (heat stress), fuel poverty, usability and the need for greater resilience.

Participation
BPE should mean mutual cooperation for all involved in the process – *with* inhabitants rather than *for* them. Co-designing BPE protocols and standards for a housing project with future (newbuild) or present (retrofit) inhabitants, followed by co-learning, can develop significant user-centred design insights about how homes work and how to improve the performance. Involving the house builder in this co-design process ensures that a virtuous learning circle evolves from the inhabitant feedback. Inhabitants as 'citizen scientists' can make effective performance evaluations of their own homes through increasingly cheap sensors and display technologies that track temperature, humidity, and water and energy use discreetly and accessibly. Increasing provision of 'how to' information on the web and finding new ways to archive and mine BPE data can also help with participation. Harnessing all this nationally can begin to build a broader epidemiology of housing performance that is long overdue.

Prototyping
Each housing development is embedded in a unique confluence of climate, place, culture and resource availability. Offsite construction offers a promising way forward, providing it responds to these factors. There is a major opportunity here to build in BPE testing of these types of housing solutions on specific sites to ensure good performance through the fine-tuning of design and construction. There is also an opportunity for prototype homes to be tested 'live' through demonstration sites such as the BRE Innovation Park, Scottish Housing Expo or international building exhibitions. This already happens with the highly successful international Solar Decathlon student competition.

Education
Learning through the lens of BPE is vital for designers to understand that there is much more to a home than simply producing a predictive model. BPE skills, knowledge and understanding must become mandatory professional validation criteria for architecture and engineering in every country. Some programmes already teach BPE using a case study approach, but many do not. If BPE is embedded in the early years, it is possible to break the tutors' habit of thinking that it is a 'bolt-on' activity, separate from the studio work. Embedding BPE into education and practice in this way makes it intrinsic to the core mission of designing for performance, and reduces the many factors that contribute to the 'performance gap'. Mandatory CPD can keep practitioners up to date with BPE developments in a rapidly changing field. Professional and educational institutions need to take a lead on all of this, to avoid their marginalisation by other commercial organisations with an interest in this field.

Closing the loop

How good is good enough?
BPE sense-checks all stages of any housing project. However, it is important not to overdo BPE activities, as this can disengage inhabitants and clients alike, as Julia Green explains:

'… they wanted go in there and cover everything in kit and do it for 10 years and take data every 5 seconds. We just can't cope with that. It's not feasible or practical for a customer perspective. And the sort of data we were being hit with was totally unmanageable.'

This book has recommended a 'light-touch' BPE study as an introductory step. How far should housing BPE go after this? This depends on the issues identified. Focusing on just five 'killer variables' can be an effective way to use BPE:

- Comfort
- Responsiveness
- Ventilation
- Layout
- Communicated design intent

These few variables account for the vast majority of design variance in offices.[7] It may well be that these are also the same 'killer variables' for housing. In any case, BPE should be always been undertaken on a 'need to know' basis, only drilling down as necessary.

Socio-cultural factors

Knowing inhabitants' past housing conditions and practices can help anticipate 'unexpected behaviours' such as energy use, which frequently confound design assumptions and predictions.[8] This is possible in retrofit projects, when the inhabitants remain in place, but not for new homes where designers need to anticipate inhabitants' practices as best they can. Engaging with local inhabitants at the start of any BPE process is best, rather than designers relying on their own experience, which can easily omit a host of design considerations for a particular place and culture. Understanding housing design and performance as continuously 'negotiated' territory between designers, managers and inhabitants[9] is a far more effective way of deploying BPE.

Emergent problems

Emergent problems need to be identified to close the performance gap more effectively. A BPE study can do this by synthesising discreet cultural, economic, environmental, social and technical factors. Housing BPE is heuristic and aims to develop a rich picture of what is happening in a home, using active listening and responding. Different housing performance problems will emerge as societies change, which must be taken into account by BPE methodology. The changes we are seeing just now – with an increasingly extreme climate, and coastlines sinking into the rising sea – demand that BPE methods evaluate how well homes can adapt to the uncertain future. BPE can use up-to-date predicted weather data, rather than past weather records, to make recommendations that take account of this. Homes also need to cope better with the replacement of obsolete technologies as new ones emerge, or old ones wear out; evaluating and promoting resilience/flexibility can be a key part of BPE to address this need.[10]

Setting the agenda: 10 questions

Housing performance clearly has to improve. Alarmingly, the UK is underperforming against three-quarters of the UN's Sustainable Development Goals (SDGs) set for 2030, and many housing organisations have still not seriously embedded SDGs into their policies and procedures.[11] Ten urgent questions for policymakers need to be answered to address these issues – and some key solutions are put forward below.

How do you intend to verify housing performance?

At the moment, housing quality is generally verified in terms of estimated rather than actual performance. Governments plan to improve housing quality based on estimates in Energy Performance Certificates (EPCs). Homes routinely underperform by 100% in terms of predicted energy use, which shows just how impoverished this idea of improving housing on the basis of estimated performance really is. Policymakers need to legislate for the evaluation of housing performance in reality, and define a verification process within the building regulations. As the CCC states: 'More focus should be placed on monitoring and reporting actual rather than modelled performance ...'[12] As Jon Bootland points out:

'Even if you don't change the building regulation targets, if we could get buildings to perform to their actual existing targets, that's a huge shift. On energy, ventilation and overheating – those three things at least.'

Current, not new, housing technologies have to reduce climate change effects. This is because the average time for commercialising new technologies is around 40 years, and global carbon emissions must be drastically reduced in the next 11 years to avoid catastrophic climate change.[13] Policymakers therefore need to focus more on verifying the actual performance of existing technologies.

How do you intend to engage the inhabitant?

The introduction of smart meters has so far only produced around 3% savings in the UK, yet the UK government's ambitious Clean Growth Strategy believes a smart systems plan could unlock savings of up to £40 billion to 2050. But how 'smart' is this really for inhabitants, who can actually feel very disempowered by these systems?[14] Using smart systems to balance the power grid puts the energy suppliers in charge of deciding when cheap energy is available, and the inhabitants must adjust their behaviour accordingly. BPE should test the appropriateness of this approach against actual reductions in energy demand, and social acceptability (see Figure 15.3).

Figure 15.3 Just how helpful are new 'smart' systems for homes? BPE can test this with inhabitants

How, then, will future housing design promote inhabitant engagement? We will certainly see more robotic interfaces and virtual displays in the home. But is this enough? The agenda for promoting genuine interactive and adaptive learning between the physical home and its inhabitant needs more development.[15] BPE can evaluate the interactive adaptivity of home interfaces to make the process of designing and constructing housing more appropriate for inhabitants.

How will you fully embed BPE into construction?

Offsite construction combined with BIM offers an amazing opportunity to embed BPE within a more effective housing design and delivery process. Research is attempting to improve the modelling of housing energy demand and carbon emissions, by validating this using data from previous monitoring studies.[16] However, these models are often still based on data gathered from traditional monitoring practices, which are themselves flawed in terms of representativeness.[17] They could easily repeat the same mistakes as before by not paying enough attention to individual design contexts. How can this be avoided?

By building simple monitoring sensors into the housing fabric during the production stage and drawing on inhabitant feedback, it is possible to richly evaluate housing performance *and* inhabitant practices, feeding these directly back into the design and modelling process through a BIM facilities management object. The BPE data links directly to product details, as well as giving inhabitants the same data for them to act on. In this way, homes can self-heal and develop through automated BPE that reports performance and helps to improve product design. However, there are clearly ethical issues to address here, which are considered next.

How will you scale up housing BPE, ethically?

BPE apps make it possible for inhabitants to report and evaluate the conditions in their home. The question is, how can this data be gathered in a way that does not exploit the inhabitant and gives them control over their privacy in relation to third parties? There is a real need to safeguard people against data exploitation by internet platforms and the internet of things. Policymakers need to develop BPE collaboration protocols with suitably robust ethics procedures for such platforms; these should consider how the data is used, whether it is aggregated or disaggregated, what granularity of data is shared and who it is shared with, as well as allowing inhabitants to opt out of data collection/data sharing at any time.

There is also a strong ethical argument for inhabitants having total control of their data. Any data gathering should also form a seamless part of inhabitants' everyday living habits, rather than asking them to do additional work. This means windows with inbuilt sensors to monitor their opening; discreet room sensors to monitor temperature, humidity and CO_2; and energy sensors in cookers, washing machines, water systems, heating systems, lighting systems and so on. Crucially, inhabitants would have to be given the ability to switch off these sensors at any time to preserve their privacy. Policymakers urgently need to clarify this BPE area in terms of responsibilities.

How will you ensure that homes are environmentally sound?

LCA is not usually part of housing BPE studies or undertaken separately by house builders in the UK. Policymakers and practitioners need to ensure the measurement of overall resource consumption as operational energy demand diminishes and homes use a greater proportion of embodied energy to achieve this reduced demand.[18] Although there are sectoral limits for CO_2e in the UK, further work is needed to translate these into specific benchmarks and targets for operational and embodied greenhouse gases (GHGs) for homes. There is also the wider issue of how the materials in our homes affect the natural environment overall. The RIBA is developing a sustainability overlay to its POE guidance that can be built into the BPE process. Is there the will from policymakers to develop centres of research excellence in housing LCA to enable this approach to BPE in the future?

How will you develop appropriate evaluation criteria for housing BPE?

Measuring housing performance is primarily based around energy efficiency and thermal comfort. More attention needs to be given to measuring indoor air quality, ventilation, lighting and noise levels, as vital feedback for human health and wellbeing. The regulatory steady-state modelling and standards used to predict housing performance also need improving. The adoption of the Passivhaus PHPP package would be a vast improvement on the current SAP and RdSAP procedures currently used in the UK. Greater improvement will happen once BPE is routine and data is continuously fed back to refine such modelling. An even better modelling process would factor in climate predictions. But perhaps the most radical step needed is for housing BPE to incorporate feedback based on how people actually choose to live in their homes – as a form of 'practice' feedback which considers whole routines as a context for any measurements.[19]

Why don't we have proper housing MOT checks?

The 2017 fire in the Grenfell Tower in London was the worst UK housing disaster since the Second World War. The resulting Hackett Report rightly led to a ban on combustible external walling material for high-rise buildings over a certain height. One reason so many people died in this fire was because fire and maintenance responsibilities were not well enforced, with a failure to regularly check for compliance. Inhabitants need to know what changes have been made to their home in the past, and that their home is healthy, safe and working well – both for their benefit and the planet's wellbeing. This is why cars have MOTs – to keep people safe and reduce pollution.

Any proposed building 'green passports' currently planned by policymakers[20] need, like MOTs, to be based on actual performance, with clear lines of responsibility for who measures what, and who should respond to the results. Integrated building regulations are essential here, combining annual checks on ventilation and thermal systems, energy feedback from smart meters, as well as regular fabric assessments. Without this reality check, inhabitants, clients and practitioners will be still be 'flying blind' in terms of trying to understand how well their homes are performing. This checking process would also help to ensure adequate maintenance of the home.

How should BPE be regulated?

BPE should be mandatory, if we want to improve housing design and performance. We are beyond waiting for voluntary measures to take effect, given that after 40 years of BPE in practice, only 3% of chartered architects actually undertake this activity in the UK, despite the routine use of BPE demonstrating significant improvement in design practice and performance.

The CCC 2018 Progress Report states that the UK government should:
'Strengthen compliance and enforcement framework so that it is outcomes-based, places risk with those able to control it, provides transparent information and a clear audit trail, with effective oversight and sanctions.'[21]

It goes on to say:

'Fully integrating real-world performance into the metrics, certification and standards framework for new and existing buildings offers opportunities to shift mind-sets and incentives towards actual performance (in place of a compliance-focused approach), narrow the performance gap, improve enforcement and reduce energy bills and carbon emissions.'[22]

This integration of BPE is needed for housing around the world, now. The question remains – can the delivery of this be left to the regulators and existing organisations, given their poor performance to date, or should a new independent and national Institute of Building Performance be established to oversee this? The current contradictions between the BS standards for BIM/Soft Landings and other POE standards, in terms of what methods to use and for how long, demonstrate how dangerous it would be to demand a prescriptive performance-based approach to BPE methodology. Policymakers need to address this question also.

How can we tackle climate change for real in the housing sector?

The housing sector in the UK significantly missed the target of a 31% reduction in GHG emissions in 2017 compared to 1990 levels.[23] To keep within a 1.5°C global temperature rise means reducing per-capita carbon emissions by *at least* 86–96% below 1990 levels in the UK by 2050.[24] This is particularly challenging given that the sector is growing, some homes are getting larger, and inhabitants are also choosing higher comfort levels for the same financial cost when moving to new homes, or following a retrofit. Meanwhile insulation for existing homes is one of the cheapest carbon-cutting measures available, but the cancellation of UK government incentives has caused a 95% drop since 2012.[25] The CCC has called for four actions by the government to put emissions reductions on track:

1. Support the simple, low-cost options.

2. Commit to effective regulation and strict enforcement.

3. End the chopping and changing of policy.

4. Act now to keep long-term options open.

This is another agenda for demanding housing BPE, because without it there is no guarantee that the installation of simple low-cost options will work. Thermal imaging, for example, reveals whether insulation has been poorly installed, which it often is. A study for the Northern Ireland Housing Executive in 2014 revealed that just 9% of homes fitted with cavity wall insulation were actually fit for purpose[26]. This type of underperformance plays havoc with planned reductions in emissions from the housing sector, undermines the credibility of the industry, and again highlights the urgent need for robust regulation in relation to BPE, compliance and checking.

Can builders and retrofitters provide a real performance guarantee?

BPE can help develop new housing products and services as well as improve existing ones, via a contractual guarantee process, as illustrated by the Energiesprong programme (see Chapter 10). It is now possible for every home to have guaranteed performance levels (and costs) in terms of energy – at least for the actual fabric and services. This would still allow inhabitants to choose to use (and pay for) energy over and above the basic performance guarantee where necessary. It could also tackle the level of carbon emissions associated with the energy supply. This demand-led approach would reconfigure how the supply side thinks and acts. It would shift housing design culture from design intentions, to designing to perform.

As a next step, housing BPE can also help to define other guaranteed levels of service to be provided related to needs – indoor air temperatures, humidity, hot water usage per day and so on. Setting these levels would create new social norms to help with adjusting inhabitant practices to work within appropriate resource use limits needed for current, and future, sustainability. This may seem unpalatable, but given that the current resource use in the UK is about three times more than what is sustainable, something has to be done.

The primer in the next section of this book provides a tool for taking housing BPE forward.

A final word on housing BPE

This book has attempted to sketch out a brief history of housing BPE, followed by a more extensive examination of the values and the developing theory behind it, as well as explaining the methods for doing it. The agenda looks to a future where designing to perform, based on BPE feedback, will be ubiquitous and routine. But we are far from that, just now. We have just 11 years left in which to reduce our carbon emissions across the board by at least 50% in 2030, to avoid a climate catastrophe. The temptation is to make energy use and carbon emissions the focus of BPE, but that would be a mistake.

Housing BPE needs to broaden its purpose. It needs to address the need for homes to be liveable, robust, adaptable and resilient in the long term – not just in terms of adjusting to cope with climate changes such as increasing temperature extremes, storms, flooding, fires and drought, but also to show how well homes accommodate people's needs throughout their lives. BPE needs to include values, methods and standards, which assess inclusive design as well as resilience. This means evaluating the degree to which homes can be adjusted for increased disability in later life, as well as providing immediate access for the disabled.

BPE has been promoted on the basis of improved energy efficiency, profitability for house builders, contractors and design teams, reputation (more sales and rentals), inhabitant satisfaction (fewer complaints) and build quality (fewer defects). But this is not a sustainable model for promoting BPE, as explained by Elizabeth Shove, a sociologist from Lancaster University:

'Programmes of energy efficiency are politically uncontroversial *precisely because they take current interpretations of "service" for granted* ... The unreflexive pursuit of efficiency is problematic not because it doesn't work or because the benefits are absorbed elsewhere, as the rebound effect suggests, but because it does work ... to sustain, perhaps escalate, but never undermine ... *increasingly energy intensive ways of life.*'[27] (Author's italics)

The clear message here is that BPE should always help with the process of designing and constructing homes which are appropriate, and which take account of these wider issues. Effective housing BPE feedback should ultimately question the purpose of house building in terms of how people want to live and how resource use is framed within broader societal forces which may be undermining the needs of people and their global home. BPE must question design intentions as well as evaluate them. And, more than ever, we need the courage and commitment to do this, for all concerned.

PRIMER:
HOW TO DO HOUSING BPE

This standalone primer for housing BPE outlines the various actions that should be undertaken to ensure an effective outcome. It provides clients, the design team and educators with a route through the rationale, set-up, process and communication framework needed to establish a participatory learning cycle in practice. It gives practical information on how to get started with a 'light-touch' BPE investigatory study, progressing to diagnostic and forensic levels, and works within the UK regulatory framework.

Key BPE principles, methods and techniques are signposted with references for further information. Required actions are set out under each of the headings below:

- Rationale and principles of housing BPE
- Engagement and ethics: briefing stage actions
- Communication
- BPE procedure and methods
- Making a difference

The primer can also be read in tandem with the previous book chapters, which provide further information and are signposted in each section of the primer.

Rationale and principles of housing BPE (see Chapters 2, 3, 4)

Housing BPE is fundamentally about learning to improve performance through continuous evaluation reflection and feedback, at all stages of the housing life cycle (see Figure 16.1).

The top seven reasons for housing clients and design teams to carry out BPE as a routine activity are as follows:

- Process and product improvement
- Reduced risk and hidden liabilities
- Reduced defects
- Reduced maintenance
- Futureproofing
- Customer satisfaction
- Better reputation

All of the above result in increased profitability. This is an effective way to introduce BPE to new clients and signal its benefits to all stakeholders, as part of embedding BPE in practice.

There are seven cyclical principles to adhere to when carrying out a housing BPE study (see Figure 16.2):

- Engage at the beginning
- Ensure good ethics
- Set up good communication
- Start with a light-touch BPE
- Drill down where necessary
- Aim to make a difference
- Embed the learning process

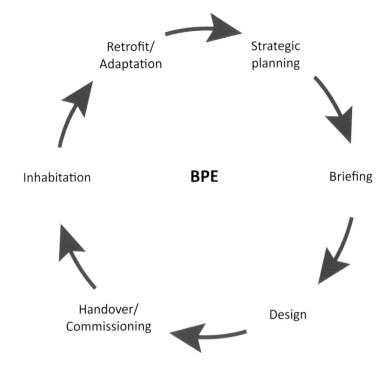

Figure 16.1 The housing life cycle

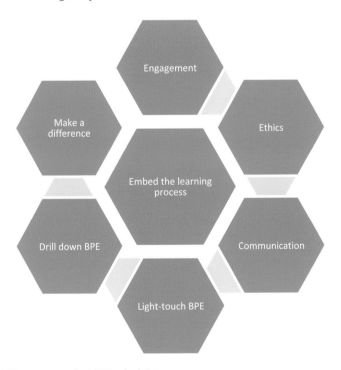

Figure 16.2 The seven cyclical BPE principles

The next sections of the primer set out what to do in relation to these principles sequentially, starting with the initial client engagement.

Engagement and ethics: briefing stage actions (see Chapters 12,13,14)

Introduce housing BPE to the team

Aim to introduce BPE as an aspiration at the start of any internal organisational review by either the client organisation or design team. This provides maximum leverage to ensure that BPE is embedded within an organisation's ethos, policy, procedures and learning culture. Alternatively, BPE can be introduced via any specific housing project.

Explain the BPE benefits at the initial project feasibility meeting in order to gain acceptance for the idea. Engage with the highest management level possible within a client organisation, so that the BPE process is communicated with authority to others. Provide an evidenced-based 30-minute BPE presentation. Show evidence from previous studies where BPE has improved performance to convince the team that it is worth engaging with. Use telling photos, charts and diagrams. Describe the light-touch BPE process and methods. Mention the possibility of needing to drill down if issues are identified, using additional methods.

Define the scope and outcomes

A clearly defined scope and anticipated outcomes of a housing BPE study should ensure accountability between all parties involved, either through a contract or by exchange of letters. This should occur at the briefing stage, following the initial discussion at the feasibility stage. The scope depends on the following:

- Funding available
- Type of housing project
- Purpose of the BPE study
- Level of investigation desired
- Anticipated outcomes

Align the BPE methods and outcomes with the agreed design intentions and the selected UN Sustainable Development Goals (SDGs) below in order to fulfil the following basic housing needs (see Figure 16.3):

- Health and wellbeing
- Clean water and sanitation
- Affordable and clean energy
- Sustainable communities
- Sustainable consumption and production
- Climate action
- Biodiversity

Set up benchmarks and modelling

Define the sustainable design standards and benchmarks against which to evaluate the housing performance. This establishes the BPE study baseline against which to gauge performance and plan for improvement. Current international and national standards include the following:

- LEED[12]
- LBC[3]
- WELL[4]
- BRE Home Quality Mark (UK only)[5]

Benchmarks directly related to basic energy in-use targets include the following:
- SAP (UK building regulation standard[6] – energy and carbon)
- Passivhaus (international standard[7] – energy only)

Ethics procedure (see Chapter 13)

Include all key stakeholders in the preparation, undertaking and reporting of the BPE study, including the housing developer, inhabitants, design team and contractor. Ensure the BPE team has suitable interdisciplinary skills to develop an ethical BPE procedure. If not, consult with a local academic institution to develop this.

Prepare the following essential light-touch BPE documents, which must have ethical approval where BPE activities directly affect the inhabitants:
- Contract
- Briefing document
- Project information form and consent form for: home tour, questionnaire, and energy/water data collection/audit and thermographic imaging

Figure 16.3 BPE should encompass key UN Sustainable Development Goals, including biodiversity

- Inhabitant questionnaire
- Interview guide and questions for post-visit discussion
- Home tour guide

Further information and approved guidance will be needed for a more diagnostic study, including an approved inhabitant interview guide and questions, and an approved survey and monitoring procedure.

Agree on data-sharing procedures from the outset, and make sure everyone is aware of how the results will be publicly disseminated. In sensitive cases, a short report can be provided for external consumption, and a more detailed report for internal use only.

Contract (see Chapter 14)

Specify the housing BPE project in an appendix to the main housing contract general terms and conditions. An exchange of letters, with an enclosed contract for mutual signing between the relevant parties (client and BPE team), will suffice for a light-touch study. The appendix should summarise the BPE project brief, specifying all tasks, timescales and milestones to be achieved. A Gantt chart is a good way to capture these (see Figure 16.4). This contract should be agreed between all parties, including inhabitant representatives where these exist.

Check the BPE contract conditions with a legal expert, to be on the safe side. Where there is more than one client organisation involved in the study, a Memorandum of Understanding should be signed between all parties, with one client agreeing to act as the lead party, to avoid confusion.

Communication (see Chapter 14)

Set up an appropriate and effective communication strategy to maximise the benefits of the BPE process. Identify a key representative for each party involved and agree meetings between them in relation to the project milestones. For a 'light-touch' study carried out over just a few months, simply organise an introductory meeting to describe the process to the representatives and a final meeting to present project outcomes. Discuss any recommendations and agree the wider dissemination of the findings, with further meetings planned only if necessary.

For a longer 'diagnostic' study it is important to have regular progress meetings and milestones every two to three months. Regular contact in between these formal meetings is useful to address any matters arising.

Keep in touch with the project inhabitants and encourage their participation by inviting them to an early meeting on site, advertised to all inhabitants via social media and/or by individual letters via the client. Describe the BPE process and provide a question and answer session. Give a presentation to all inhabitants once the project has finished. This helps with the learning process for all concerned.

Feed back findings and recommendations into practice policy and procedures for the design team and client to inform future projects, with the aim of making BPE a routine activity within each organisation (see 'Organisational learning over time' in Chapter 6 and Chapter 14 and p 206). Communicate the findings more widely to other relevant organisations and, if possible, share the datasets to build up BPE databases (see 'Wider learning and impact' in Chapter 14 and p 206).

BPE procedure and methods (Chapters 5, 6, 7, 8)

Light-touch BPE study
Four basic BPE questions need to be addressed, initially:

1. Is the home (or homes) physically performing as expected?

2. Are the inhabitants' needs met in their home(s)?

3. Are there any physical and/or social problems that need solving just now?

4. How can we improve our housing for the future?

A light-touch BPE study can address these questions by using the following six activities in sequence:

1. Review of key design, construction and specification documents related to the project intentions
Examine the outline brief and specification documents and compare these to the key drawings for a typical home in the project (1:100/1:50 layout plans, long and cross-sections, all elevations). Check key 1:5 details around openings, foundation and roof junctions, as built. Highlight any areas of potential concern compared to best practice and climate change predictions. Note any changes to the specifications or drawings. This should take no more than half a day (design team).

2. Basic energy and water use audit
Obtain permission from inhabitants to use their data anonymously using an ethics procedure (see Chapter 13). Gather energy and water meter data for individual homes via the client or service suppliers involved. Break down the annual figures (KWh and litres) in relation to the relevant net floor area to compare with your chosen benchmarks (KWh/m^2/pa) (litres/home/pa). Where this is not possible, use annual water and energy bills and compare with benchmarks to see if these are excessive or not. If the results are worse than the benchmarks, use further BPE methods to find out why (see under 'Diagnostic BPE', below) (design team).

3. Thermographic survey
Obtain permission to carry out a brief survey inside and outside of at least one home, using a basic thermal imaging camera to identify any obvious heat-loss issues. The survey should take around half a day, and then a further half-day to analyse and report on. Highlight any areas of concern to the client. Use someone trained in thermography, as it is easy to get wrong.[8] Use BS EN 13187:1999 Thermal performance of buildings. Qualitative detection of thermal irregularities in building envelopes. Infrared method.

4. Simple survey of the inhabitants

Use the domestic Building Use Studies (BUS) questionnaire[9] or equivalent (see Chapter 7). Aim for a 50–100% return rate from all homes in the project, using either hand delivery and pick-up of the questionnaire or digital means. Hand delivery is usually more successful. It could take around a couple of days to deliver and input the data, depending on the number of homes involved. The initial analysis included in the cost of the BUS survey will highlight any issues. Do the survey at least one year after initial inhabitation to capture seasonal variations. Pay close attention to the individual inhabitant comments. These may well highlight new areas of concern or even best practice. A completed questionnaire means the inhabitant has given permission to use their data anonymously (design team).

5. Short tour of one home

Obtain permission from the inhabitant to do a home tour, allowing an hour for the visit with another hour for the team discussion afterwards (see Chapters 7 and 8). This should happen after the above surveys have been analysed, as this highlights issues for the home tour to focus on. The group should be no larger than five to seven people for the walkthrough, including the inhabitants, and a representative from the client, architect, engineer and contractor. Use a home tour guide to undertake this activity in sequence, which should cover all rooms and the exterior. Compare findings to the initial document review (design team).

6. Spot-check environmental conditions

Use a simple hand-held multi-function 4-in-1 environmental meter to correlate temperature, humidity, light and sound levels with any initial observations made during the walkthrough, and highlight any issues. Take readings in the main living room and the main bedroom as key areas. These meters are relatively cheap and worthwhile investing in for the practice (trained personnel).

The results should be analysed, triangulated, discussed and presented in a final report (see Chapter 14 and p 205). If there are any areas of underperformance with an unidentifiable cause, advise the client to move to the next level of diagnostic investigation.

Checklist for an initial light-touch BPE project

RIBA Work Stage	BPE action	Guidance
0: Strategic Definition	• Review previous BPE studies • Initial BPE meeting and contract	• Explain BPE, agree BPE project • Develop 30-minute presentation • Give top 10 reasons for doing BPE • Use seven principles of BPE
1: Preparation/ Brief	• Define scope of BPE project • Agree contract • Set up ethics and communications	• Align brief with 7 SDGs • Define brief, targets, benchmarks, modelling, BPE methods, ethics, expected outcomes • Use expertise (academic/consultant) to develop ethics procedure • Involve the inhabitants

2–6: Concept Design to Handover	• Utilise feedback from previous BPE studies	• Use organisational learning in practice via electronic templates for BPE to improve work at each stage of contract
7: In Use	• Do BPE study using six activities: • Document review • Energy and water audit • Thermographic survey • Inhabitant survey • Home tour • Environmental spot-check	• Undertake these activities in sequence to get maximum benefit from the comparative analysis at each stage and between stages • Analyse what is happening and why things work well, or present challenges • Use reporting structure set out in this primer • Disseminate findings and data widely to client, inhabitants, practice and beyond for learning and impact • Embed learning in practice

Diagnostic BPE study

Use the methods below sequentially as necessary to diagnose any problems identified in the light-touch BPE study and to bring any further problems in the housing project to the surface.

Full document review

Examine the project strategies and standards documentation carefully (see chapter 7), and compare it to the brief, as-built project drawings, specifications and construction details to see if they conform to the following:

- Health and safety strategy and standards
- Spatial quality/layout strategy and standards
- Construction strategy and standards
- Environmental systems and 'smart' control/metering strategy and standards
- Maintenance and operational strategy and standards
- Sustainability strategy and standards
- Handover and inhabitant induction strategy and standards

If any of these are missing, this should be noted for action. Check if the documented strategies address climate change adaptation. If not, the project will most likely fail in the future (design team).

Photographic survey

Develop a full photographic survey of the construction externally and internally (see Chapter 7). Use the photo and initial thermographic survey to compare with relevant structural and construction details as part of the construction audit and diagnose any faults (design team).

Construction audit

Undertake a detailed inspection of the housing fabric via a walkthrough, noting anything unusual compared to the construction strategy and standards audit (see Chapter 7) (design team).

BPE Workload Plan and programme

Month · 2019-20	Mar	Apr	May	Jun
1. BPE project implementation				
1.1 Orientation and detailed planning of project				
1.2 Contractual agreement		M1		
1.3 Ethics application			M2	
2. Construction audit				
2.1 Construction process review				M3
2.2 Data capture of drawings, specifications - comparison				
2.3 Design team interviewed in relation to design intentions				
2.4 Photo survey of construction fabric				
2.5 Evening publicity meeting with inhabitants (questions)				
2.6 Inhabitant agreements				
3. Post-occupancy evaluation				
3.1 Thermal imaging audit of construction fabric				
3.2 Thermal imaging report				
3.3 Airtightness testing (others)				
3.4 Comfort and control questionnaire x all homes				
3.5 Installation and commissioning processes checked				
3.6 Home user guide evaluated				
3.7 Handover processes evaluated				
3.8 Feedback/recommendations to developer (initial)				
3.9 Portable monitoring kit ordered and installed				
3.10 Monitoring of homes for one year (energy, water, IAQ)				
3.11 Home tour /interviews with sample inhabitants				
3.12 SAP and EPC calculation / energy model (others)				
3.13 Thermal bridging calculations / heat flux tests (others)				
4. Analysis				
4.1 Semi-structured interviews/home tour				
4.2 Questionnaire data input + analysis				
4.3 Monitoring analysis				
4.4 Triangulation of all results				
5. Reporting				
5.1 Interim report				
5.2 Final report (M8)				
6. Dissemination and impact				
6.1. Internal learning for developer (presentation + guidance)				
6.4. Presentations to inhabitants - learning process				

Figure 16.4 Example of a BPE project Gantt chart

Jul	Aug	Sep	Oct	Nov	Dec	Jan	Feb	Mar	Apr	May	Jun	Jul	Aug	Sep	Oct	Nov	Dec

M4

M5

M6

M7

M8

M8

BS 7543:2015 Guide to durability of buildings and building elements, products and components.

Re-test air permeability

Employ a specialist to re-test air permeability levels compared to the original certification for building regulations (see Figure 16.5). Include smoke tests to diagnose air-leakage. Note any deviations and identify sources. Provide recommendations for remedying any faults identified.

Re-commission the services

Employ an engineer to identify and re-test relevant commissioning processes (see Chapter 7). This should encompass all service controls and systems, including all fixed renewable energy, heating, lighting, water and ventilation appliances. Check against any relevant commissioning plan, manuals, standards, drawings and specifications. Check all control settings against manufacturers' recommendations. Visually inspect for any obvious errors in installation, faulty equipment or damage. Test all heating, cooling and ventilation systems to ensure they are working properly and are fully adjusted. Carry out air-flow measurements to test mechanical ventilation systems. Check guidance for the inhabitant on how to use these systems, and note anything missing. Recommend interventions where needed.

BS EN 15251:2007 Indoor environmental input parameters for design and assessment of energy performance of buildings addressing indoor air quality, thermal environment, lighting and acoustics (under review)

Handover review

Identify any induction standards in place for the project. Observe the induction meetings for inhabitants, including before, during and after handover (see Chapter 7). Check personnel and routines involved, as well as any written guidance provided. Triangulate this with the organisational requirements, inhabitant survey comments and any interview analysis, to highlight where the induction needs improving and/or inhabitants need further training (design team).

Energy/Carbon calculation review

Obtain from the relevant design team representative the original predicted energy/carbon emissions calculations (e.g. SAP or RdSAP) and any associated certification. Re-work the calculations to check them. Report any errors and the calculated impact on the predicted performance levels (design team).

Heat-loss measurement through key elements

Use temperature probes fixed to key wall, floor and roof areas to understand where heat loss is occurring in the external fabric of the housing and the degree to which it is occurring. Always combine this method with an airtightness test and thermal imaging survey in order to properly diagnose the results (see Chapter 7) (expert).

BS ISO 9869-2:2018 Thermal insulation. Building elements. In-situ measurement of thermal resistance and thermal transmittance. Infrared method for frame structure dwelling

BS EN 13187:1999 Thermal performance of buildings. Qualitative detection of thermal irregularities in building envelopes. Infrared method

Energy, water and environmental monitoring

Employ experts to carry out any monitoring and analysis (see Chapter 7). Fit energy and water sub-meters to key home sub-circuits (heating, cooker, washing machine, renewable energy) to identify whether the underperformance is due to faulty equipment, installation or commissioning, or inhabitant routines. Ensure that energy meters can record both imported and exported electricity to the National Grid in order to isolate renewable energy performance issues.

Continuously monitor temperature, humidity, lighting, acoustics and CO_2 levels over a minimum period of around 12 months (including peak summer and winter temperature periods) and triangulate against related issues raised in the light-touch BPE study. A 12-month period will account for all seasonal variations and related inhabitant practices. A two-year period from handover is needed to account for a one-year living adjustment process (including defects made good) followed by one year of normalised conditions.

Figure 16.5 A diagnostic airtightness test can quickly identify if seals have failed over time

Calibrate and place a single multi-sensor at body height (measuring at least temperature, humidity and CO_2 levels) securely in the living room and main bedroom as a minimum. Place additional sensors in the bathroom and kitchen if there are also issues identified in these areas. Include acoustic and lighting sensors if these have been raised as issues. Avoid placing them on external walls or near heat sources and direct sunlight. Sensors should be standalone with downloadable data collected regularly (which also provides the opportunity to visit homes and make regular observations) or linked via wired circuits to a remote central data collector. Wireless sensor systems sometimes fail due to signal interference and need to be tested very carefully. Obtain weather data from the nearest local weather station for comparison of internal and external environmental conditions. Alternatively, buy a mini-weather station linked by wire or wirelessly to a data collector.

BS EN 15251:2007 Indoor environmental input parameters for design and assessment of energy performance of buildings addressing indoor air quality, thermal environment, lighting and acoustics (under review)

BS EN ISO 16000-1:2006 Indoor air. General aspects of sampling strategy

ISO 7726, Ergonomics of the thermal environment. Instruments for measuring physical quantities

BS EN 13032-4:2015 Light and lighting. Measurement and presentation of photometric data of lamps and luminaires. LED lamps, modules and luminaires

BS EN ISO 16283-2:2018 Acoustics. Field measurement of sound insulation in buildings and of building elements. Impact sound insulation

Interviews

Employ an expert to carry out semi-structured interviews with a representative demographic sample of inhabitants using predefined questions, based on key issues raised in the inhabitant questionnaire (see Chapters 4, 7 and 13). Useful guidance can be found in *The Sage Handbook of Interview Research: The Complexity of the Craft,* 2012. It is also useful to interview client, contractor and design team representatives. Allow 30–45 minutes maximum for each interview. Record and fully transcribe the interview – allow 4–6 hours for each interview for this process. Do not rely on interview notes only – this will result in bias and miss crucial details. Code and analyse the interview(s) using appropriate methods (expert).

Finally, analyse and compare findings from any or all of the above BPE methods against the design intentions as manifest in the full document review to reveal any disparities. Communicate any urgent findings as they emerge to the client directly. Triangulate between all methods to reveal any hidden issues and/or additional explanations. Discuss all findings in the final report. If there are any unresolved issues or areas of underperformance, where the cause cannot be identified, advise the client to move to the next level of forensic investigation.

Checklist for additional work on diagnostic BPE project

RIBA Work Stage	BPE action	Guidance
0: Strategic Definition	• Review previous light-touch BPE study • Initial BPE meeting and contract	• Identify and agree additional BPE work needed with client, design team and inhabitants
1: Preparation/ Brief	• Define scope of additional BPE project • Agree contract • Set up ethics and communications	• Define additional BPE methods needed • Define expected outcomes • Define additional brief, targets, benchmarks, modelling, BPE methods, ethics, expected outcomes • Use expertise (academic/ consultant) to develop additional ethics procedure
6: Handover	• Full document review • Photographic survey • Construction audit • Re-test air permeability • Re-commission services • Handover review	• Undertake these additional activities, ideally at the commissioning and handover stage, to fine-tune the housing performance and improve inhabitant understanding • Undertake activities sequentially for maximum benefit when comparing results
7: In Use	• Energy/CO_2 calculations review • Heat-loss measurement • Monitoring • Interviews	• Undertake the energy calculations review ahead of any monitoring to establish the correct design targets against which to benchmark monitoring results • Undertake interviews after the physical monitoring, to enable more targeted interview questions • Carry out analysis, reporting and dissemination as for the light-touch study

Forensic BPE Study

Forensic BPE methods provide the deepest level of investigation necessary to address persistent and unresolved housing performance issues. Discuss and agree proposed forensic work with the client in terms of additional costs, commitments and timescales, set against the possible benefits. Careful judgement is needed here to ensure maximum benefit in relation to any extra costs. Always employ relevant experts to carry out any of the activities below.

Co-heating test

A co-heating test[10] measures the overall heat loss through the fabric of a home. It is a complex and costly procedure carried out by experts, and needs extensive permission, but it does provide a strong basis for understanding fabric performance in relation to other factors (see Chapter 7).

Moisture movement

Undertake a whole-house approach to understand the pathology of moisture movement.[11] Use probes in key areas to identify any moisture movement through the fabric in relation to identified issues, for example in walls, floors and ceilings, and to highlight emergent issues.

BS 6576:2005+A1:2012 Code of practice for diagnosis of rising damp in walls of buildings and installation of chemical damp-proof courses

BRE Digest 245 Rising Damp in Walls – diagnosis and treatment, 2007

Extended monitoring

Obtain permission to install a comprehensive weather station unit on site, measuring external temperature, wind speed, solar gain and humidity as a minimum. Diagnose seasonal microclimate effects on housing performance. Carry out external measurements of toxicity levels where necessary. Employ an expert to measure indoor toxicity levels related to CO, VOCs, NO_2, ozone and formaldehyde as well as particulates of dust (PM2.5 and PM10) or other allergens. This involves sophisticated techniques and equipment to meet international standards. Alternatively, use low-cost pollution sensors via mobile phone apps for initial rough measurements (see Chapter 7).[12]

BS EN ISO 16000-5:2007 Indoor air. Sampling strategy for volatile organic compounds (VOCs).

Usability survey

Obtain permission to ask inhabitants to complete a usability survey[13] of their environmental controls and services. This task can be done by the design team. This works best when combined with other BPE methods (see Chapter 4 and 7), such as a home tour, inhabitant interviews, induction evaluation and document review to reveal any underlying causes.

Occupancy patterns

Obtain permission to identify when the home is occupied and what vents, doors and windows are being opened, using passive infrared PIR motion sensors positioned over all exterior doors, and binary sensors fixed to all window openings. This can help you understand how and why heating, lighting and ventilation systems are/are not used.

Ethnographic studies

Obtain permission to carry out observations of inhabitant routines, and encourage inhabitants to keep diaries or logging of key events. Take short videos during a home tour of key issues and/or encourage inhabitants to video their routines in relation to key issues such as ventilation, heating and cooling strategies. Use these studies to diagnose inhabitant practices in the home in relation to any identified issues and to bring hidden issues to the surface (see Chapter 4 and 7).[14]

Thermal comfort survey

Obtain permission for a thermal comfort survey to reveal whether or not inhabitants are actually comfortable in their homes, as opposed to what they might say. Inhabitants complete a comfort survey within the home each day at regular intervals over key periods of time in winter and summer. A fixed or portable anemometer measured air velocity, with sensors for measuring wet-bulb radiant temperatures and humidity. Use experts to carry out this labour-intensive exercise.

BS EN ISO 7730:2005. Ergonomics of the thermal environment. Analytical determination and interpretation of thermal comfort using calculation of the PMV and PPD indices and local thermal comfort criteria

Focus groups

Focus groups[15] are good for digging into difficult issues and gaining a 360-degree perspective. Obtain permission and include the design team, housing developer and/or inhabitant representatives. Between six and twelve people is ideal for the agreed focus topic. Design a focus group guide to be used by the facilitator, ideally on site. Allow two hours for the activity itself, and 12 hours for transcribing the recording. Avoid taking notes for the same reason given for interviews. This work is best done by experts (see Chapter 4 and 7).

Checklist for additional forensic BPE project

RIBA Work Stage	BPE action	Guidance
0: Strategic Definition	• Review previous diagnostic BPE study • Initial BPE meeting and contract	• Identify and agree additional BPE work needed with client, design team and inhabitants
1: Preparation/Brief	• Define scope of forensic BPE project • Agree contract • Set up ethics and communications	• Define additional BPE methods needed • Define expected outcomes • Define additional brief, targets, benchmarks, modelling, BPE methods, ethics, expected outcomes • Use expertise (academic/consultant) to develop additional ethics procedure
7: In Use	Methods: • Co-heating test • Moisture movement analysis • Extended monitoring • Usability survey • Occupancy patterns • Ethnographic studies • Thermal comfort survey • Focus groups	• Use any or all of these methods as necessary to understand issues still unresolved from the diagnostic BPE study • Carry out analysis, reporting and dissemination as for the light-touch study • Provide separate appendices for each method and results as part of the final report

Description and location				
Greenwood Filters within MVHR unit				
Usability criteria	**Poor**			**Excellent**
Clarity of purpose				
Intuitive switching				
Labelling and annotation				
Ease of use				
Indication of system response				
Degree of fine control				

Comments

No labelling whatsoever – if you are not familiar with the unit you don't know they are there. The filters are easy to take out and clean but some units clash with the door frames. There is little indication of when they are getting clogged up – a warning light might be useful? The whole front panel must come out to access the heat exchanger filter every two years – this operation in some units on site is very difficult due to clashes with doorframes and very poor location of units high up in narrow spaces.

A usability survey forensically evaluates the ergonomics of service controls in a home

Making a difference

Design the analysis and reporting of the BPE studies to have maximum impact on the project and organisational learning, such that immediate and strategic action is taken where necessary to make a positive difference (see Chapter 6, 10 and 11).

Analysis

Analyse what is happening. The design team can do this for the light-touch BPE level, but experts should carry out this work, at diagnostic or forensic levels in BPE. Compare the BPE results with the key design intentions and targets, including energy use, carbon dioxide emissions, and indoor environmental quality. Benchmark the results against other sustainable design standards and housing studies by drawing on existing databases, or by comparison with similar housing BPE project reports. Align the comparison with the overall key performance indicators for the practice/client organisation to help improve the performance of the homes, and relate it to other case studies and key lessons (see Chapter 5).

Analyse why things work well, or present challenges. Compare the key findings from the document analysis, physical surveys and measurements, and social surveys. Identify any issues in terms of not meeting targets and best practice. It is particularly important initially to cross-check drawings, specifications, strategies and standards against the physical construction and performance of the homes. This establishes a baseline against which to subsequently evaluate any social findings. However, BPE analysis also works reflexively. The findings from social surveys can often be used to explain why certain physical situations are occurring, such as damp, overheating in summer, underheating in winter, poor indoor air quality, lighting or acoustics. Be systematic and cross-check all physical findings with all social findings, as this can also unearth hidden and unknown issues that can be caught early before becoming major problems.

Compare context-rich quotations from residents on key aspects of performance with other findings from the BPE study, particularly when these draw attention to specific processes and design details in the home. This can generate deep learning in terms of organisational and design processes. Identify issues, lessons and recommendations in relation to the original design intentions.

Do not share raw or processed data more widely within or beyond the participating organisations unless there is explicit ethical permission given to do so (see Chapter 13).

Reporting

Provide a short final report of no more than about 20 pages with appendices attached. A more in-depth project needs an interim report and a longer final report prepared by an expert team. The interim report should signpost any significant or unusual initial trends for the client to act on if necessary, while the final report should provide detailed analysis for the client, design team and contractor as well as an overview aimed at a wider non-specialist audience.

The main body of a final report needs to contain the following key sections:

- Executive summary
- Introduction and details of the housing project
- Aims and objectives of the BPE project
- BPE methods used
- Basic results
- Analysed and evaluated findings
- Key lessons and recommendations related to remedial actions and improving the housing process

Create a simple table which identifies priority actions and the investment needed, for the immediate (now), medium (one to two years) and long (five years +) term. This can effectively mobilise the housing project team and client to improve their practices. Avoid pages and pages of recommendations, which are less likely to be read.

Celebrate positive results first and identify critical remedial actions last. Any political sensitivity over the BPE findings should be taken into account, with a separate report for public consumption produced if necessary. Additional sub-reports on particular BPE activities can make the reporting more resilient and agile, enabling the client and BPE team to choose which sub-reports to send to different entities as required. All BPE reports should support the findings with suitably distilled and appropriate data, maps, graphs, diagrams, tables, photographs and sketches as necessary (see Chapter 14).

Organisational learning

To embed and improve organisational learning take the following steps:

1. Define the BPE study
2. Describe the situation and context
3. Collect evaluative data and analyse it
4. Review the data and highlight good practice as well as any issues
5. Tackle the issues by introducing changes
6. Monitor the changes
7. Analyse evaluation about the changes
8. Review the changes and decide what to do next[16]

For steps 6 to 8, develop a simple electronic template to record key lessons from the BPE final report, to store and disseminate at relevant organisational levels. Make sure this completed template, as well as the BPE final report, is accessible for everyone in the design practice and/or housing organisation as part of a quality management system. Tag the templated findings into the organisational briefing and BIM processes, key products and design details, particularly where performance issues and solutions have been highlighted. This systematic reporting needs to be reinforced with regular BPE workshops, reflective team reviews and site visits to embed the learning in practice. Use the BPE findings to inform further practice-based research and development (see Chapter 6 ,14 and 15).

Wider learning and impact

Generate wider learning and impact from your BPE studies through office-sponsored seminars. The Good Homes Alliance and other NGOs can help organise these in the UK. Generate policy impact through relevant professional bodies and government agencies. Target cross-party parliamentary groups of MPs who aim to improve the quality and performance of the built environment. Share the BPE studies and benchmarking with key professional institutions to generate more robust national BPE datasets for housing performance, which are currently missing. The analysis from housing BPE studies can also inform practice-based design guidelines for future housing projects (see Chapter 14).

Turn your BPE studies into guidance for industries across the built environment to help establish your reputation for housing BPE. Share the findings through CPD events, conferences, practice pamphlets, books and research journal papers co-authored with academics. Offer BPE projects as case studies for academics and students. Use social media to disseminate housing BPE findings and gain useful feedback. Take care when dealing with the news media, however, as reporters can tend to want to slant the BPE findings towards their particular agenda. Ask for a copy of any proposed text and check it. Always obtain permission from the client before undertaking any publicity of BPE findings, unless this has been given previously (see Chapter 13).

This primer provides a guide to getting started with housing BPE – feel free to share it with others.

References

Introduction

1. K.B. Janda, 'Buildings don't use energy: people do', *Architectural Science Review*, 54, 2011, pp 15–22.
2. D.A. Schön, *The Reflective Practitioner: How Professionals Think in Action*, New York, Basic Books, 1983.
3. J. Till, *Architecture Depends*, Cambridge, Massachusetts, MIT, 2009.

Chapter 1

1. P. Vitruvius, *The Ten Books on Architecture*, New York, Dover, 1960.
2. A. Tinniswood, *By Permission of Heaven: The Story of the Great Fire of London*, London, Jonathan Cape, 2003.
3. M.J. Daunton, 'Health and housing in Victorian London', *Medical History*, 126, 1991.
4. A.S. Wohl, *The Eternal Slum: Housing and Social Policy in Victorian London*, London, Edward Arnold, 1977, p 148.
5. Ibid.
6. C. Porteous, *The New Eco-architecture: Alternatives from the Modern Movement*, London, Spon, 2002.
7. R. Courtney, 'Building Research Establishment past, present and future', *Building Research & Information*, 25, 1997, pp 285–91.
8. T. Bedford, C.G. Warner & F.A. Chrenko, 'Observations on the natural ventilation of dwellings', *Journal of the Royal Institute of British Architects*, 1943.
9. R. McCutcheon, 'Technical change and social need: the case of high rise flats', *Research Policy*, 4, 1975, pp 262–89.
10. M. Hollow, 'Governmentality on the Park Hill estate: the rationality of public housing', *Urban History*, 37, 2010, pp 117–35.
11. M. Power, *Domestic Fuel Policy: Report by the Fuel and Power Advisory Council*, 1946, p 22.
12. R. Courtney, 1997.
13. D. Watt, *Building Pathology: Principles and Practice*, Oxford, Blackwell, 2007, p 91.
14. R. Courtney, 1997.
15. T. Tsubaki, 'Planners and the public: British popular opinion on housing during the second world war', *Contemporary British History*, 14, 2000, pp 81–98.
16. Mass Observation, 'Some psychological factors in home building', *Town and Country Planning*, vol. 11, no. 41, 1943, p 9.
17. K. Ellsworth-Krebs, L. Reid & C.J. Hunter, 'Home-ing in on domestic energy research: "House", "home", and the importance of ontology', *Energy Research & Social Science*, 6, 2015, pp 100–8.
18. Mass Observation, 'An enquiry into people's homes', *The Advertising Service Guild*, 1943.
19. T. Tsubaki, 2000.
20. R. Courtney, 1997.
21. T. Tsubaki, 2000.
22. R. McCutcheon, 1975.
23. Ibid., p 284.
24. M. Hollow, 2010 (p 129, citing SLS/HLG 331.8335F, CSHC, Park Hill Redevelopment, 1960).
25. Ibid.
26. E.T. Hall, *The Hidden Dimension: Man's Use of Space in Public and Private*, London, Bodley Head, 1969.
27. D.V. Canter, *Psychology for Architects*, London, Applied Science Publishers, 1974.
28. D.V. Canter, et al., *Psychology and the Built Environment*, London, Architectural Press, 1974.
29. H. Sanoff, 'Techniques of Evaluation for Designers: With the Assistance of Gary Coates, C.T. Jackson, Jr., and J. Thompson', Research Laboratory Monologue, North Carolina State University, 1968.
30. A. Friedmann, C. Zimring & E. Zube, *Environmental Design Evaluation*, New York, Springer Science + Business Media, 1978.
31. 'Post-occupancy evaluation' was first used as a term in in 1975 by Herbert McLaughlin of KMD Architects in the USA. (Preiser, Wolfgang F.E., and Nasar, Jack, 2008, 'Assessing building performance: its evolution from post-occupancy evaluation', in *ArchNet-IJAR: International Journal of Architectural Research*, vol. 2, issue 10).
32. M. Kantrowitz, 'Has Environment and Behavior Research Made a Difference?', *Environment and Behavior*, 17, 1985, pp 25–46.
33. J. Zeisel, *Inquiry by Design: Tools for Environment-Behaviour Research*, Cambridge University Press, 1984.
34. C. Cooper Marcus & W. Sarkissian, *Housing as if People Mattered: Site Design Guidelines for Medium-density Family Housing*, Berkeley and Los Angeles, California, University of California Press, 1986.
35. W.F.E. Preiser, H.Z. Rabinowitz & E.T. White, *Post-occupancy Evaluation*, Abingdon, Routledge, 1988.
36. M. Kantrowitz, 1985.
37. J. Dahir, *Communities for Better Living: Citizen Achievement in Organization, Design and Development*, New York, Harper and Brothers, 1950, p 9.
38. R. McCutcheon, 1975, p 288.
39. H. Sanoff, *Integrating*

Programming, Evaluation and Participation in Design: A Theory Z Approach (Routledge Revivals), Abingdon, Routledge, 1992, p vii.

40. P. Jenkins & L. Forsyth, *Architecture, Participation, and Society*, London, Routledge, 2010.

41. H. Sanoff, 'The social implications of residential environments', *International Journal of Environmental Studies*, 2, 1971, pp 13–19.

42. N. Wates, *Community Architecture: How People are Creating Their Own Environment*, London, Penguin, 1987.

43. P.B. Jones, D. Petrescu & J. Till, *Architecture and Participation*, London, Spon, 2004.

44. P. Jenkins & L. Forsyth, 2010.

45. J.F.C. Turner, *Freedom to Build: Dweller Control of the Housing Process*, New York, Macmillan, 1972.

46. C. Ward, *Housing: An Anarchist Approach*, London, Freedom Press, 1976.

47. J. Broome, *The Self-build Book: How to Enjoy Designing and Building Your Own Home]*, Bideford, Green Books, 1991.

48. P. Jenkins & L. Forsyth, 2010.

49. Ibid.

50. F. Stevenson & N. Williams, *Sustainable Housing Design Guide for Scotland*, Norwich, Stationery Office, 2000.

51. F. Stevenson, & N. Williams, 'Sustainable Housing Design Guide for Scotland', web archive, 2007, https://www.webarchive.org.uk/wayback/archive/20170105043652/http://www.gov.scot/Topics/Built-Environment/Housing/investment/shdg/ (accessed 19th December 2018).

52. W.F.E. Preiser, 'Post-occupancy evaluation: how to make buildings work better', *Facilities*, 13, 1995, pp 19–28.

53. W.F.E. Preiser & J. Vischer. *Assessing building performance*, Oxford, Elsevier Butterworth-Heinemann, 2005.

54. R. Gupta, M. Gregg, S. Passmore & G. Stevens, 'Intent and outcomes from the Retrofit for the Future programme: key lessons', *Building Research & Information*, 2015, pp 1–18.

55. T. Lees & M. Sexton, 'An evolutionary innovation perspective on the selection of low and zero-carbon technologies in new housing', *Building Research & Information*, 42, 2014, pp 276–87.

56. Ecological Design Group, 'Evaluation of Kincardine O'Neil innovative rural housing design project Canmore Place', *Research from Communities Scotland*, Edinburgh, 87, 2005.

57. B. Bordass & A. Leaman, 'Making feedback and post-occupancy evaluation routine 1: A portfolio of feedback techniques', *Building Research & Information*, 33, 2005, pp 347–52.

58. R. Lowe, L.F. Chiu & T. Oreszczyn, 'Socio-technical case study method in building performance evaluation', *Building Research & Information*, 2017, pp 1–16.

59. A. Leaman, F. Stevenson & B. Bordass, 'Building evaluation: practice and principles', *Building Research & Information*, 38, 2010, pp 564–77.

60. E. Shove, 'What is wrong with energy efficiency?', *Building Research & Information*, 2017, pp 1–11.

61. B. Bordass & A. Leaman, 'A new professionalism: remedy or fantasy?', *Building Research & Information*, 41, 2013, pp 1–7.

Chapter 2

1. S.K. Allen, et al., 'Climate change 2013: the physical science basis. Contribution of working group 1 to the fifth assessment report of the Intergovernmental Panel on Climate Change', IPCC, Cambridge, UK and New York, USA, 2013a.

2. M. Jennings, N. Hirst & A. Gambhir, 'Reduction of carbon dioxide emissions in the global building sector to 2050', Imperial College, Grantham Institute for Climate Change, 2011.

3. EC, *Greenhouse Gas Emissions by Economic Activity and by Pollutant*, EU-28, European Commission, 2014.

4. J. Palmer, D. Godoy-Shimizu, A. Tillson & I. Mawditt, 'Building performance evaluation programme: Findings from domestic projects – making reality match design', 38, Swindon, Innovate UK 2016.

5. IPCC, 'Intergovernmental Panel on Climate Change fifth assessment report (AR5)', 2013b.

6. E. Shove, 'What is wrong with energy efficiency?', *Building Research & Information*, 2017, pp 1–11.

7. https://www.statista.com/statistics/827267/average-household-water-usage-united-kingdom-uk/

8. https://www.mcginley.co.uk/news/how-to-make-a-glass-of-water/bp142/

9. Z.M. Gill, M.J. Tierney, I.M. Pegg & N. Allan, 'Measured energy and water performance of an aspiring low energy/carbon affordable housing site in the UK', *Energy & Buildings*, 43, 2011, pp 117–25.

10. S. Sturgis, *Targeting Zero: Whole Life and Embodied Carbon Strategies for Design Professionals*, London, RIBA Publishing, 2017.

11. J. Giesekam, J. Barrett, P. Taylor & A. Owen, 'The greenhouse gas emissions and mitigation options for materials used in UK construction', *Energy & Buildings*, 2014.

12. C. De Wolf, F. Pomponi & A. Moncaster, 'Measuring embodied carbon dioxide equivalent of buildings: A review and critique of current industry practice', *Energy & Buildings*, 140, 2017, pp 68–80.
13. M. Sinclair, J. Toole, M. Malawaraarachchi & K. Leder, 'Household greywater use practices in Melbourne, Australia', *Water Science & Technology: Water Supply*, 13, 2013, pp 294–301.
14. P. Mang & B. Haggard, *Regenerative Development and Design: A Framework for Evolving Sustainability*, New Jersey, John Wiley and Sons, 2016.
15. N. Wang, et al., 'Ten questions concerning future buildings beyond zero energy and carbon neutrality', *Building and Environment*, 119, 2017b, pp 169–82.
16. I. Hamilton, A. Summerfield, T. Oreszczyn & P. Ruyssevelt, 'Using epidemiological methods in energy and buildings research to achieve carbon emission targets', *Energy & Buildings*, 154, 2017, pp 188–97.
17. https://ec.europa.eu/clima/policies/strategies_en0
18. https://ec.europa.eu/energy/intelligent/projects/en/projects/zebra2020
19. R. Lowe, 'Technical options and strategies for decarbonizing UK housing', *Building Research & Information*, 35, 2007, pp 412–25.
20. http://www.ukcip.org.uk/glossary/
21. J.J. Porter, S. Dessai & E.L. Tompkins, 'What do we know about UK household adaptation to climate change? A systematic review', *Climatic Change*, 127, 2014, pp 127–371.
22. M. Baborska-Narożny, F. Stevenson & M. Grudzinska, 'Overheating in retrofitted flats: Occupant practices,

learning and interventions', *Building Research and Information*, 45, 2017, pp 40–59.
23. K. Williams, et al., 'Retrofitting England's suburbs to adapt to climate change', *Building Research & Information*, 41, 2013, p. 517-531
24. F. Stevenson, M. Baborska-Narożny & P. Chatterton, 'Resilience, redundancy and low-carbon living: Co producing individual and community learning', *Building Research & Information*, 44, 2016, pp 789–803.
25. B. Rodríguez-Soria, J. Domínguez-Hernández, J.M. Pérez-Bella & J.J. del Coz-Díaz, 'Review of international regulations governing the thermal insulation requirements of residential buildings and the harmonization of envelope energy loss', *Renewable & Sustainable Energy Reviews*, 34, 2014, pp 78–9.
26. W. Pan & H. Garmston, 'Compliance with building energy regulations for new-build dwellings', *Energy*, 48, 2012, pp 11–22.
27. E. Burman, D. Mumovic & J. Kimpian, 'Towards measurement and verification of energy performance under the framework of the European directive on the energy performance of buildings', *Energy*, 77, 2014, pp 153–63.
28. D. Jenkins, S. Simpson & A. Peacock, 'Investigating the consistency and quality of EPC ratings and assessments', *Energy*, 138, 2017, pp 480–9.
29. A. Stone, D. Shipworth, P. Biddulph & T. Oreszczyn, 'Key factors determining the energy rating of existing English houses', *Building Research & Information*, 42, 2014, p 725–738
30. J. Palmer, D. Godoy-Shimizu, A. Tillson & I. Mawditt, 2016.

31. W. Pan & H. Garmston, 2012.
32. https://www.gov.uk/government/publications/ventilation-approved-document-f
33. C. M. Wang, et al., 'Unexpectedly high concentrations of monoterpenes in a study of UK homes', *Environmental Science: Processes & Impacts*, 19, 2017a, pp 528–37.
34. J.F. Nicol, & M. Wilson, 'A critique of European Standard EN 15251: Strengths, weaknesses and lessons for future standards', *Building Research & Information*, 39, 2011, pp 183–93.
35. N. Dodd, M. Cordella, M. Traverso & S. Donatello, 'Level(s) – A common EU framework of core sustainability indicators for office and residential buildings: Part 3: How to make performance assessments using Level(s) (Draft Beta v1.0)', *JRC Technical Reports*, 210. Brussels, European Commission, 2017.
36. C. Colin Nugent, et al., 'Noise in Europe', *European Environment Agency*, Luxembourg, 2014.
37. H. Notley, et al., *National Noise Attitude Survey 2012 (NNAS2012) Summary Report*, Department for Environment and Rural Affairs, 2014.
38. N. Wang, et al., 2017b.
39. R. Kaplan & R.H. Matsuoka, 'People needs in the urban landscape: Analysis of landscape and urban planning contributions', *Landscape and Urban Planning*, 84, 2008, pp 7–19.
40. M. Baborska-Narożny, F. Stevenson & P. Chatterton, 'Temperature in housing: Stratification and contextual factors', *Engineering Sustainability*, 2016, pp 125–37.
41. CABE, 'What it's like to live there: the views

of residents on the design of new housing', London, *Commission for Architecture and the Built Environment*, 2005.

42. R. Simmons, 'Constraints on evidence-based policy: Insights from government practices', *Building Research & Information*, 2015, pp 1–13.

Chapter 3

1. D. King, *Engineering a Low Carbon Built Environment: The Discipline of Building Engineering Physics*, London, The Royal Academy of Engineering, 2010.
2. S.V. Szokolay, *Introduction to Architectural Science: The Basis of Sustainable Design*, London, Routledge, 2014.
3. Ibid.
4. Ibid.
5. Ibid.
6. M. Baborska-Narożny, F. Stevenson & P. Chatterton, 'Temperature in housing: Stratification and contextual factors', *Engineering Sustainability*, 169, 2016, pp 125–37.
7. S.V. Szokolay, 2014.
8. M. Zinzi & E. Carnielo, 'Impact of urban temperatures on energy performance and thermal comfort in residential buildings: The case of Rome, Italy', *Energy & Buildings*, 2017.
9. S.V. Szokolay, 2014.
10. Ibid.
11. H. Chappells, 'Comfort, well-being and the socio-technical dynamics of everyday life', *Building Research & Information*, 2, 2010, pp 286–98.
12. P.O. Fanger, *Thermal Comfort: Analysis and Applications in Environmental Engineering*, Copenhagen, Danish Technical Press, 1970.
13. F. Nicol, M. Humphries & S. Roaf, *Adaptive Thermal Comfort: Principles and*

Practice, Abingdon, New York: Earthscan, 2012.

14. M.A. Humphreys, *Adaptive Thermal Comfort: Foundations and Analysis*, London, Routledge, 2016.
15. F. Nicol & S. Roaf, 'Post-occupancy evaluation and field studies of thermal comfort', *Building Research & Information*, 33, 2005, pp 338–46.
16. C. Schweizer, et al., 'Indoor time-microenvironment-activity patterns in seven regions of Europe', *Journal of Exposure Science and Environmental Epidemiology*, 17, 2007, pp 170–81.
17. G. McGill, L. O. Oyedele & K. McAllister, 'Case study investigation of indoor air quality in mechanically ventilated and naturally ventilated UK social housing', *International Journal of Sustainable Built Environment*, 4, 2015, p 58–77
18. ISO, '16814:2008(E) building environment design – indoor air quality – methods of expressing the quality of indoor air for human occupancy', International Organization for Standardization, Geneva, 2008.
19. S. Vardoulakis, et al., 'Impact of climate change on the domestic indoor environment and associated health risks in the UK', *Environment International*, 85, 2015, pp 299–313.
20. WHO, *WHO Guidelines for Indoor Air Quality: Selected Pollutants*, Bonn, 2010.
21. S. Vardoulakis, et al., 2015.
22. P. Sassi, 'Thermal comfort and indoor air quality in super-insulated housing with natural and decentralized ventilation systems in the south of the UK', *Architectural Science Review*, 60, 2017, pp 167–79.
23. A. Lewis, 'Daylighting in older people's housing: Barriers to compliance

with current UK guidance', *Lighting Research & Technology*, 47, 2015, pp 976–92.

24. J. Swanson, 'Childhood cancer in relation to distance from high-voltage power lines in England and Wales: A case-control study', *Journal of Radiological Protection: Official Journal of the Society for Radiological Protection*, 25, 2005, p 336.
25. S.J. Genuis, 'Fielding a current idea: exploring the public health impact of electromagnetic radiation', *Public Health*, 122, 2008, pp 113–24.
26. S. Darby, et al., 'Radon in homes and risk of lung cancer: Collaborative analysis of individual data from 13 European case-control studies', *BMJ*, 330, 2005, pp 223–6.
27. A.M. Makantasi & A. Mavrogianni, 'Adaptation of London's social housing to climate change through retrofit: A holistic evaluation approach', *Advances in Building Energy Research*, 2015, pp 1–26.
28. M. Mulville & S. Stravoravdis, 'The impact of regulations on overheating risk in dwellings', *Building Research & Information*, 2016, pp 1–15.
29. R. Gupta & M. Gregg, 'Preventing the overheating of English suburban homes in a warming climate', *Building Research & Information*, 41, 2013, pp 281–300.
30. S. Vardoulakis, et al., 2015.
31. F. Stevenson, I. Carmona-Andreu & M. Hancock, 'The usability of control interfaces in low-carbon housing', *Architectural Science Review*, 56, 2013, pp 70–82.

Chapter 4

1. B. Pilkington, R. Roach & J. Perkins, 'Relative benefits of technology

and occupant behaviour in moving towards a more energy efficient, sustainable housing paradigm', *Energy Policy*, 39, 2011, pp 4962–70.

2. B.K. Janda, 'Buildings don't use energy: People do', *Architectural Science Review,* 54, 2011, pp 15–22.

3. H. Jarvis, 'Saving space, sharing time: Integrated infrastructures of daily life in cohousing', *Environment and Planning A*, 43, 2011, pp 560–77.

4. Z.M. Gill, M.J. Tierney, I.M. Pegg & N. Allan, 'Low-energy dwellings: The contribution of behaviours to actual performance', *Building Research & Information*, 38, 2010, pp 491–508.

5. E. Shove, *Comfort, Cleanliness and Convenience: The Social Organization of Normality*, Oxford, Berg Publishers, 2003.

6. N.J. Moore, V. Haines & D. Lilley, 'Improving the installation of renewable heating technology in UK social housing properties through user centred design', 2015.

7. S. Pink, K. Mackley, V. Mitchell, M. Hanratty, C. Escobar-Tello, T. Bhamra & R. Morosanu, 'Applying the lens of sensory ethnography to sustainable HCI', *ACM Transactions on Computer-Human Interaction (TOCHI)*, 20, 2013, pp 1–18.

8. E. Shove, 'What is wrong with energy efficiency?', *Building Research & Information*, 2017, pp 1–11.

9. E. Shove, 'Beyond the ABC: Climate change policy and theories of social change', *Environment and Planning A*, 42, 2010, pp 1273–85.

10. R. Cole & R. Lorch, *Buildings, Culture and Environment: Informing Local and Global Practices*, Oxford, Blackwell, 2003.

11. G. Walker, 'The dynamics of energy demand: Change, rhythm and synchronicity', *Energy Research & Social Science*, 1, 2014, pp 49–55.

12. L. Kuijer & M. Watson, '"That's when we started using the living room": Lessons from a local history of domestic heating in the United Kingdom', *Energy Research & Social Science*, 28, 2017, pp 77–85.

13. V. Fabi, R.V. Andersen, S. Corgnati & B.W. Olesen, 'Occupants' window opening behaviour: A literature review of factors influencing occupant behaviour and models', *Building and Environment*, 58, 2012, pp 188–98.

14. H. Erhorn, 'Influence of meteorological conditions on inhabitants' behaviour in dwellings with mechanical ventilation', *Energy & Buildings*, 11, 1988, pp 267–75.

15. F. Stevenson & M. Baborska-Narozny, 'Housing performance evaluation: Challenges for international knowledge exchange', *Building Research & Information*, 2017, pp 1–12.

16. J. Love, A.C.G. Cooper, M. Shipworth & M. Ucci, 'From social and technical to socio-technical: Designing integrated research on domestic energy use', *Indoor and Built Environment*, 24, 2015, pp 986–98.

17. S.W. Aboelela, E. Larson, S. Bakken, O. Carrasquillo, A. Formicola, S.A. Glied, J. Haas & K.M. Gebbie, 'Defining interdisciplinary research: Conclusions from a critical review of the literature', *Health Services Research*, 42, 2007, pp 329–46.

18. J. Love, et al., 2015.

19. F. Stevenson & H.B. Rijal, 'Developing occupancy feedback from a prototype to improve housing

production', *Building Research & Information*, 38, 2010, pp 549–63.

20. T. Crosbie & K. Baker, 'Energy-efficiency interventions in housing: Learning from the inhabitants', *Building Research & Information*, 38, 2010, pp 70–9.

21. S. Pink, et al., 2015.

22. C. Shrubsole, A. Macmillan, M. Davies, N. May & M. Ucci, '100 Unintended consequences of policies to improve the energy efficiency of the UK housing stock', *Indoor and Built Environment*, 23, 2014, pp 340–52.

23. J. Chen, J.E. Taylor & H.-H. Wei, 'Modeling building occupant network energy consumption decision-making: The interplay between network structure and conservation', *Energy & Buildings*, 47, 2012, pp 515–24.

24. H. Putra, C. Andrews & J. Senick, 'An agent- based model of building occupant behavior during load shedding', *An International Journal*, 10, 2017, pp 845–59.

25. C. Robson, *Real World Research: A Resource for Users of Social Research Methods in Applied Settings*, Chichester, Wiley, 2011.

26. E. Shove, 2004.

27. C. Grandclément, A. Karvonen & S. Guy, 'Negotiating comfort in low energy housing: The politics of intermediation', *Energy Policy*, 84, 2015, pp 213–22.

28. E. Shove, 2010.

29. A.H. Maslow, *Motivation and Personality*, New York, Harper, 1954.

30. S. Thielke, M. Harniss, H. Thompson, S. Patel, G. Demiris & K. Johnson, 'Maslow's Hierarchy of Human Needs and the adoption of health-related technologies for older adults', *Ageing

International, 37, 2012, pp 470–88.

31. M. Max-Neef, *Human Scale Development: Conception, Application and Further Reflections*, New York, USA and London, Apex Press, 1991.

32. https:// sustainabledevelopment. un.org/sdgs

33. J. Gibson, *The Ecological Approach to Visual Perception*, Boston, Houghton Mifflin Company, 1979.

34. T. Ingold, 'Culture and the perception of the environment', *Bush Base: Forest Farm – Culture, Environment and Development*, 1992.

35. H. Heft, 'Affordances and the body: An intentional analysis of Gibson's ecological approach to visual perception', *Journal for the Theory of Social Behaviour*, 19, 1989, pp 1–30.

36. D.A. Norman, *The Design of Everyday Things*, London, MIT Press, 1998.

37. M. Baborska-Narożny & F. Stevenson, 'Service controls interfaces in housing: Usability and engagement tool development', *Building Research & Information*, 47, 2019, pp 290–304.

38. J.D. Gould & C. Lewis, 'Designing for usability: Key principles and what designers think', *Communications of the ACM*, 28, 1985, pp 300–11.

39. J.D. Gould & C. Lewis, 1985.

40. R. Gupta & S. Chandiwala, 'Understanding occupants: Feedback techniques for large-scale low-carbon domestic refurbishments', *Building Research & Information*, 38, 2010, pp 530–48.

41. F. Stevenson & H.B. Rijal, 'Developing occupancy feedback from a prototype to improve housing production', *Building Research & Information*, 38, 2010, pp 549–63.

42. J.D. Gould & C. Lewis, 1985.

43. J. Zeisel, *Inquiry by Design: Tools for Environment-Behaviour Research*, Cambridge, Cambridge University Press, 1984.

44. C. Tweed, 'Socio-technical issues in dwelling retrofit', *Building Research & Information*, 41, 2013, pp 551–62.

45. M. Heidegger, *Building, Dwelling, Thinking*, New York, Harper and Row, 1971.

46. C. Tweed & G. Zapata-Lancaster, 'Interdisciplinary perspectives on building thermal performance', *Building Research & Information*, 2017, pp 1–14.

47. H. Heft, 'Perceptual Information of "An Entirely Different Order": The "Cultural Environment" in The Senses Considered as Perceptual Systems', *Ecological Psychology*, 29, 2017, p. 122-145

48. F. Stevenson, I. Carmona-Andreu & M. Hancock, 'The usability of control interfaces in low-carbon housing', *Architectural Science Review*, 56, 2013, pp 70–82.

49. I. Ajzen, 'The theory of planned behaviour: Reactions and reflections', *Psychology & Health*, 26, 2011, pp 1113–27.

50. F.L. Scott, C.R. Jones & T.L. Webb, 'What do people living in deprived communities in the UK think about household energy efficiency interventions?', *Energy Policy*, 66, 2014, pp 335–49.

51. Z.M. Gill, M.J. Tierney, I.M. Pegg & N. Allan, 'Low-energy dwellings: the contribution of behaviours to actual performance', *Building Research & Information*, 38, 2010, pp 491–508.

52. B. Latour, *Reassembling the Social: An Introduction to Actor Network Theory*, Oxford, Oxford University Press, 2005.

53. Z. Frances & F. Stevenson, 'Domestic photovoltaic systems: The governance of occupant use', *Building Research & Information*, 46, 2018, pp 23–41.

54. C. Grandclément, A. Karvonen & S. Guy, 'Negotiating comfort in low energy housing: The politics of intermediation', *Energy Policy*, 84, 2015, pp 213–22.

55. J. Johnson, 'Mixing humans and nonhumans together: The sociology of a door-closer', *Social Problems*, 35, 1988, pp 298–310.

56. Ibid.

57. E. Shove, 2010.

58. T.R. Schatzki, *Social Practices: A Wittgensteinian Approach to Human Activity and the Social*, Cambridge, Cambridge University Press, 1996.

59. E. Shove, *The Dynamics of Social Practice: Everyday Life and How It Changes*, Los Angeles and London, SAGE, 2012.

60. Grandclément, et al., 2015.

61. K. Gram-Hanssen, 'Standby Consumption in Households Analyzed With a Practice Theory Approach', *Journal of Industrial Ecology*, 14, 2010, pp 150–65.

62. G. Kelly, *The Psychology of Personal Constructs, Vol. 1, A Theory of Personality*, New York, W.W. Norton, 1955.

63. L. Kuijer & M. Watson, 2017.

64. T. Crosbie & K. Baker, 2010.

65. M. Sunikka-Blank & R. Galvin, 'Irrational homeowners? How aesthetics and heritage values influence thermal retrofit decisions in the United Kingdom', *Energy Research & Social Science*, 11, 2016, pp 97–108.

66. K. Gram-Hanssen,

'Existing buildings: Users, renovations and energy policy', *Renewable Energy*, 61, 2014, pp 136–40.

67. Z. Frances & F. Stevenson, 2018.

68. V. Haines & V. Mitchell, 'A persona-based approach to domestic energy retrofit', *Building Research & Information*, 42, 2014, pp 462–76.

69. Gill, et al., 2010.

70. M. Baborska-Narożny, F. Stevenson & F. J. Ziyad, 'User learning and emerging practices in relation to innovative technologies: A case study of domestic photovoltaic systems in the UK', *Energy Research & Social Science*, 13, 2016, pp 24–37.

71. M. Denny & N. Peter, 'Exploring the attitudes-action gap in household resource consumption: Does "environmental lifestyle" segmentation align with consumer behaviour?' *Sustainability*, 5, 2013, pp 1211–33.

72. K. Gram-Hanssen, 2014.

73. D. Teli, T. Dimitriou, P.A.B. James, A.S. Bahaj, L. Ellison, A. Waggott & T. Dwyer, 'Fuel poverty-induced "prebound effect" in achieving the anticipated carbon savings from social housing retrofit', *Building Services Engineering Research & Technology*, 37, 2016, pp 176–93.

74. J. Chapman, *Emotionally Durable Design: Objects, Experiences, and Empathy*, London, Earthscan, 2005.

75. C. Shrubsole, A. Macmillan, M. Davies, N. May & M. Ucci, '100 Unintended consequences of policies to improve the energy efficiency of the UK housing stock', *Indoor and Built Environment*, 23, 2014, pp 340–52.

76. G. McGill, T. Sharpe, L. Robertson, R. Gupta & I. Mawditt, 'Meta-analysis of indoor temperatures in new-build housing', *Building Research & Information*, 45, 2017, pp 19–39.

77. M. Zinzi & E. Carnielo, 'Impact of urban temperatures on energy performance and thermal comfort in residential buildings: The case of Rome, Italy', *Energy & Buildings*, 2017.

78. M. Baborska-Narożny, E. Stirling & F. Stevenson, 'Exploring the efficacy of Facebook groups for collective occupant learning about using their homes', *American Behavioral Scientist*, 61, 2017, pp 757–73.

79. R.J. Cole, J. Robinson, Z. Brown & M. O'Shea, 'Re-contextualizing the notion of comfort', *Building Research and Information*, 36, 2008, pp 323–36.

80. L. F. Chiu, R. Lowe, R. Raslan, H. Altamirano-Medina & J. Wingfield, 'A socio-technical approach to postoccupancy evaluation: interactive adaptability in domestic retrofit', *Building Research & Information*, 2014, p 1–17

81. T. Hargreaves, C. Wilson & R. Hauxwell-Baldwin, 'Learning to live in a smart home', *Building Research and Information*, 46, 2018, pp 127–39.

82. F. Stevenson, M. Baborska-Narożny & P. Chatterton, 'Resilience, redundancy and low-carbon living: Co-producing individual and community learning', *Building Research & Information*, 44, 2016, pp 789–803.

83. Baborska-Narożny, et al., 2017.

Chapter 5

1. C. Robson, *Real World Research: A Resource for Users of Social Research Methods In Applied Settings*, Chichester, Wiley, 2011.

2. B. Zeigler, A. Muzy & E. Kofman, *Theory of Modeling and Simulation*, Orlando, USA, Academic Press, 2018.

3. S.V. Szokolay, *Introduction to Architectural Science: The Basis of Sustainable Design*, London, Routledge, 2014.

4. S. Kelly, D. Crawford-Brown & M. G. Pollitt, 'Building performance evaluation and certification in the UK: Is SAP fit for purpose?', *Renewable and Sustainable Energy Reviews*, 16, 2012, pp 6861–78.

5. S. Lewis, *PHPP Illustrated: A Designer's Companion to the Passivhaus Planning Package: Second Edition,* London, RIBA Publishing, 2017.

6. https://passipedia.org/operation/operation_and_experience/measurement_results/energy_use_measurement_results

7. D.B. Crawley, J.W. Hand, M. Kummert & B.T. Griffith, 'Contrasting the capabilities of building energy performance simulation programs', *Building and Environment*, 43, 2008, pp 661–73.

8. B. Glasgo, C. Hendrickson & I.L. Azevedo, 'Assessing the value of information in residential building simulation: Comparing simulated and actual building loads at the circuit level', *Applied Energy*, 203, 2017, pp 348–63.

9. C. Shrubsole, et al., '100 Unintended consequences of policies to improve the energy efficiency of the UK housing stock', *Indoor and Built Environment*, 23, 2014, pp 340–52.

10. C. Tweed, 'Socio-technical issues in dwelling retrofit', *Building Research & Information*, 41, 2013, pp 551–62.

11. P. Kenneth Sungho & K. Ki Pyung, 'Essential BIM input data study for housing refurbishment: Homeowners' preferences in the UK', *Buildings*, 4,

2014, pp 467–87.

12. Z. Alwan, 'BIM performance framework for the maintenance and refurbishment of housing stock', *Structural Survey*, 34, 2016, pp 242–55.

13. J. Palmer, D. Godoy-Shimizu, A. Tillson & I. Mawditt, 'Building performance evaluation programme: Findings from domestic projects – making reality match design', *Innovate UK*, 38, 2016.

14. G. McGill, T. Sharpe, L. Robertson, R. Gupta & I. Mawditt, 'Meta-analysis of indoor temperatures in new-build housing', *Building Research & Information*, 45, 2017, pp 19–39.

15. G. McGill, L.O. Oyedele & K. McAllister, 'Case study investigation of indoor air quality in mechanically ventilated and naturally ventilated UK social housing', *International Journal of Sustainable Built Environment*, 4, 2015, pp 58–77.

16. F. Seguro, 'Insights from social housing projects: Building performance evaluation meta-analysis', *National Energy Foundation*, Milton Keynes, UK, 2016.

17. F. Stevenson, I. Carmona-Andreu & M. Hancock, 'The usability of control interfaces in low-carbon housing', *Architectural Science Review*, 56, 2013, pp 70–82.

18. F. Seguro, 2016.

19. C. Robson, 2011.

20. Ibid., p 190.

21. M. Baborska-Narożny, F. Stevenson & M. Grudzinska, 'Overheating in retrofitted flats: Occupant practices, learning and interventions', *Building Research & Information*, 45, 2017, pp 40–59.

22. J. Palmer, D. Godoy-Shimizu, A. Tillson & I. Mawditt, 2016.

23. F. Stevenson, I. Carmona-Andreu & M. Hancock, 2013, p 9.

24. J. Balvers, et al., 'Mechanical ventilation in recently built Dutch homes: technical shortcomings, possibilities for improvement, perceived indoor environment and health effects', *Architectural Science Review*, 55, 2012, pp 4–14.

25. F. Stevenson & H.B. Rijal, 'Developing occupancy feedback from a prototype to improve housing production', *Building Research & Information*, 38, 2010, pp 549–63.

26. F. Stevenson, I. Carmona-Andreu & M. Hancock, 2013.

27. J.D. Khazzoom, 'Economic implications of mandated efficiency in standards for household appliances', *The Energy Journal*, 1, 1980, pp 21–40.

28. R. Galvin, *The Rebound Effect in Home Heating: A Guide for Practitioners and Policymakers*, Abingdon, Oxfordshire and New York, Routledge, 2016.

29. M. Sunikka-Blank & R. Galvin, 'Introducing the prebound effect: The gap between performance and actual energy consumption', *Building Research & Information*, 40, 2012, pp 260–73.

30. E. Shove, *Comfort, Cleanliness and Convenience: The Social Organization of Normality*, Oxford, Berg, 2004.

31. F. Stevenson & H.B. Rijal, 2010.

32. S. Finlay, et al., *The Way We Live Now: What People Need and Expect from Their Homes*, London, Royal Institute of British Architects, 2012.

33. N. Terry & J. Palmer, 'Trends in home computing and energy demand', *Building Research & Information*, 2015, pp 1–13.

34. F. Wade, M. Shipworth & R. Hitchings, 'How installers select and explain domestic heating controls', *Building Research & Information*, 45, 2017, pp 371–83.

35. C. Brown & M. Gorgolewski, 'Understanding the role of inhabitants in innovative mechanical ventilation strategies', *Building Research & Information*, 43, 2015, pp 210–21.

36. T. Hargreaves, C. Wilson & R. Hauxwell-Baldwin, 'Learning to live in a smart home', *Building Research & Information*, 46, 2018, pp 127–39.

37. M. Baborska-Narożny & F. Stevenson, 'Mechanical ventilation in housing: understanding in-use issues', *Engineering Sustainability*, 170, 2017.

38. M. Baborska-Narożny, F. Stevenson & P. Chatterton, 'Temperature in housing: Stratification and contextual factors', *Engineering Sustainability*, 179, 2016, pp 125–37.

39. M. Donn, S. Selkowitz & B. Bordass, 'The building performance sketch', *Building, Research & Information*, 40, 2012, pp 186–208.

40. B. Bordass & A. Leaman, 'Making feedback and post-occupancy evaluation routine 1: A portfolio of feedback techniques', *Building Research & Information*, 33, 2005, p 349.

41. A. Leaman, F. Stevenson & B. Bordass, 'Building evaluation: Practice and principles', *Building Research & Information*, 38, 2010, pp 564–77.

42. B. Flyvbjerg, 'Five misunderstandings about case-study research', *Qualitative Inquiry*, 12, 2006, pp 219–45.

43. I. Hamilton, et al., 'Using epidemiological methods in energy and buildings research to achieve carbon emission targets', *Energy*

and Buildings, 154, 2017, pp 188–97.

44. F. Stevenson, M. Baborska-Narożny & P. Chatterton, 'Resilience, redundancy and low-carbon living: Co-producing individual and community learning', *Building Research & Information*, 44, 2016, pp 789–803.

45. T. Williamson, V. Soebarto & A. Radford, 'Comfort and energy use in five Australian award-winning houses: Regulated, measured and perceived', *Building Research & Information*, 38, 2010, pp 509–29.

46. G. Brager, H. Zhang & E. Arens, 'Evolving opportunities for providing thermal comfort', *Building Research & Information*, 43, 2015, pp 274–87.

Chapter 6

1. S. Brand, *How Buildings Learn: What Happens After They're Built*, London, Orion Books, 1994.

2. S. Pink, et al., 2017, *Making Homes: Ethnography and Design*, London, Bloomsbury Academic, 2017.

3. F. Bartiaux, K. Gram-Hanssen, P. Fonseca, L. Ozoliņa & T.H. Christensen, 'A practice–theory approach to homeowners' energy retrofits in four European areas', *Building Research & Information*, 2014, pp 1–14.

4. R, Kitchin, *Code/Space: Software and Everyday Life*, Cambridge, MIT Press, 2011, p 174.

5. Ecological Design Group, 'Evaluation of Kincardine O'Neil innovative rural housing design project Canmore Place', *Research from Communities Scotland*, 87, Edinburgh, 2005.

6. F. Stevenson & N. Williams, *Sustainable Housing Design Guide for Scotland*, Norwich, Stationery Office, 2000.

7. F. Stevenson & N. Williams,

'Longitudinal evaluation of affordable housing projects in Scotland: Lessons for procurement, design and management of low energy features', *24th International Conference of Passive Low Energy Architecture*, Singapore, Department of Architecture, National University of Singapore, 2007, pp 728–34.

8. D.A. Kolb, *Experiential Learning: Experience as the Source of Learning and Development*, London, Prentice-Hall, 1984, p 38.

9. A. Leaman, F. Stevenson & B. Bordass, 'Building evaluation: Practice and principles', *Building Research & Information*, 38, 2010, pp 564–77.

10. F. Stevenson, et al., 2011, *Building Performance Evaluation Avante Homes: Fabric Report and Occupancy Feedback, Final Report, Domestic Buildings, Phase 1 Post Construction and Early Occupation*, Innovate UK, 2011.

11. D.A. Schön, *The reflective Practitioner: How Professionals Think in Action,* New York, Basic Books, 1983.

12. H. De Haan & J. Keesom, *What Happened to My Buildings: Learning from 30 Years of Architecture with Marlies Rohmer*, Rotterdam, Nai010 Publishers, 2016.

Chapter 7

1. A. Bendell, *Benchmarking for Competitive Advantage*, London, Pitman, 1998.

2. https://www.usgbc.org/discoverleed/certification/homes/

3. https://www.usgbc.org/discoverleed/certification/homes-midrise/

4. https://www.breeam.com/discover/technical-standards/newconstruction/

5. https://living-future.org/affordable-housing/

6. https://v2.wellcertified.com/landing

7. M. Osmani & A. Reilly, 'Feasibility of zero carbon homes in England by 2016: A house builder's perspective', *Building and Environment*, 44, 2009, pp 1917–24.

8. R. Hay, F. Samuel, K.J. Watson & S. Bradbury, 'Post-occupancy evaluation in architecture: Experiences and perspectives from UK practice', *Building Research & Information*, 2017, pp 1–13.

9. BRE, 'Home quality mark one technical manual for England, Scotland and Wales', Watford, *Building Research Establishment*, 2018.

10. W.F.E. Preiser & J.C. Vischer, *Assessing Building Performance*, London, Elsevier Butterworth-Heinemann, 2005.

11. R. Hay, et al., *Pathways to POE*. Value of Architects, University of Reading, RIBA, 2017.

12. R. Hay, F. Samuel, K.J. Watson & S. Bradbury, 2017.

13. Ibid.

14. A. Leaman, F. Stevenson & B. Bordass, 'Building evaluation: Practice and principles', *Building Research & Information*, 38, 2010, pp 564–77.

15. M. Baborska-Narożny & F. Stevenson, 'Service controls interfaces in housing: Usability and engagement tool development', *Building Research & Information*, 47, 2019, pp 290–304.

16. https://www.sciencedirect.com/science/article/pii/S037877881831716X

17. P. Sassi, 'Thermal comfort and indoor air quality in super-insulated housing with natural and decentralized ventilation systems in the south of the UK', *Architectural Science Review*, 60, 2017, pp 167–79.

18. F. Nicol & S. Roaf, 'Post-occupancy evaluation and

field studies of thermal comfort', *Building Research & Information*, 33, 2005, pp 338–46.

19. R. Gupta & S. Chandiwala, 'Understanding occupants: Feedback techniques for large-scale low-carbon domestic refurbishments', *Building Research & Information*, 38, 2010, pp 530–48.

20. F. Nicol & S. Roaf, 2005.

21. https://www.busmethodology.org.uk/

22. P. Li, T.M. Froese & G. Brager, 'Post-occupancy evaluation: State-of-the-art analysis and state-of-the-practice review', *Building and Environment*, 133, 2018, pp 187–202.

23. S. Pink, et al., *Making Homes: Ethnography and Design,* London, Bloomsbury Academic, 2017.

24. Z. Frances & F. Stevenson, 'Domestic photovoltaic systems: The governance of occupant use', *Building Research & Information*, 46, 2018, pp 23-41.

Chapter 8

1. K. Gram-Hanssen, 'Standby consumption in households analyzed with a practice theory approach', *Journal of Industrial Ecology*, 14, 2010, pp 150–65.

2. K. Barker, 'Barker review of housing supply: Securing our future housing needs', *Interim Report London*, 2003.

3. F. Stevenson & H.B. Rijal, 'Developing occupancy feedback from a prototype to improve housing production', *Building Research & Information*, 38, 2010, pp 549–63.

4. A. Marshall, R. Fitton, W. Swan, D. Farmer, D. Johnston, M. Benjaber & Y. Ji, 'Domestic building fabric performance: Closing the gap between the in situ measured and modelled performance', *Energy and*

Buildings, 150, 2017, pp 307–17.

5. C. Gaze, *Information Paper: Lessons from AIMC4 for Cost-Effective, Low-Energy, Fabric-First Housing: Part 5: As-Built Performance and Post-Occupancy Evaluation*, Watford, Building Research Establishment, 2014.

6. M. Baborska-Narożny & F. Stevenson, 'Service controls interfaces in housing: Usability and engagement tool development', *Building Research & Information*, 47, 2019, pp 290–304.

7. T. Ingold, 'Anthropology contra ethnography', *Hau: Journal of Ethnographic Theory*, 7, 2017, pp 21–6.

8. S. Pink, et al., *Making Homes: Ethnography and Design*, London, UK Bloomsbury Academic, 2017.

9. M. Baborska-Narożny & F. Stevenson, 2019.

10. M. Baborska-Narożny, F. Stevenson & M. Grudzinska, 'Overheating in retrofitted flats: Occupant practices, learning and interventions', *Building Research and Information*, 45, 2017a, pp 40–59.

11. M. Baborska-Narożny, F. Stevenson & P. Chatterton, 'A social learning tool: Barriers and opportunities for collective occupant learning in low carbon housing', *Energy Procedia*, 62, 2014, pp 492–501.

12. M. Baborska-Narożny, F. Stevenson & M. Grudzinska, 2017a.

13. S. Pink, et al., 2017.

14. S. Pink, *Doing Visual Ethnography*, London, Sage Publications Ltd, 2013.

15. W. Sperschneider & K. Bagger, 'Ethnographic fieldwork under industrial constraints: Toward design-in- context', *International Journal of Human-Computer Interaction*, 15, 2003, pp 41–50.

16. S. Pink, 2013.

17. F. Stevenson & H.B. Rijal, 'Developing occupancy feedback from a prototype to improve housing production', *Building Research & Information*, 38, 2010, pp 549–63.

18. S. Pink, 2013.

19. F. Stevenson & H.B. Rijal, 2010.

20. Ofcom, *Adults' Media Use and Attitudes Report*, London, Ofcom, 2018.

21. R.C. Dalton, S.F. Kuliga & C. Hölscher, 'POE 2.0: Exploring the potential of social media for capturing unsolicited post-occupancy evaluations', *Intelligent Buildings International*, 5, 2013, pp 162–80.

22. M. Baborska-Narożny, E. Stirling & F. Stevenson, 'Exploring the efficacy of Facebook groups for collective occupant learning about using their homes', *American Behavioral Scientist*, 61, 2017b.

23. J. Gabrys, 'Citizen sensing, air pollution and fracking: From "caring about your air" to speculative practices of evidencing harm', *The Sociological Review*, 65, 2017, pp 172–92.

24. J. Palmer, D. Godoy-Shimizu, A. Tillson & I. Mawditt, *Building Performance Evaluation Programme: Findings from Domestic Projects – Making Reality Match Design*, Swindon, Innovate UK, 2016.

25. G. McGill, et al., 'Meta-analysis of indoor temperatures in new-build housing', *Building Research & Information*, 45, 2017, pp 19–39.

26. R. Lowe, L.F. Chiu & T. Oreszczyn, 'Socio-technical case study method in building performance evaluation', *Building Research & Information*, 2017, pp 1–16.

27. T. Hargreaves, C. Wilson & R. Hauxwell-Baldwin, 'Learning to live in a smart home', *Building Research*

& *Information*, 46, 2018, pp 127–39.

28. M. Baborska-Narożny, F. Stevenson & P. Chatterton, 2014.

29. F. Stevenson, M. Baborska-Narożny & P. Chatterton, 'Resilience, redundancy and low-carbon living: Co- producing individual and community learning', *Building Research & Information*, 44, 2016, pp 789–803.

Chapter 9

1. F. Stevenson & M. Baborska-Narożny, 'Housing performance evaluation: Challenges for international knowledge exchange', *Building Research & Information,* 2017, pp 1–12.

2. E. Wenger, 'Communities of practice and social learning systems', *Organization,* vol. 7, no. 2, 2000, pp 225–46.

3. F. Samuel, 'Supporting research in practice', *The Journal of Architecture,* vol. 22, no. 1, 2017, pp 4–10.

4. E. Wenger, 'Communities of practice: Learning, meaning, and identity', New York, Cambridge University Press, 2004.

5. A. Nicolaides & D.C. McCallum, 'Inquiry in action for leadership in turbulent times', *Journal of Transformative Education*, vol. 11, no. 4, 2013, pp 246–60.

6. L. Pasquale, M. Hancock & F. Stevenson, 'Embedding building performance evaluation in a medium-sized architectural practice: A Soft Landings approach', *Conference Proceedings of the 27th International Conference on Passive and Low Energy Architecture*, Louvain, Belgium, 2011.

7. J. Hines & C. Thoua, 'Post-occupancy evaluation of five schools by Architype', *Architects' Journal*, 2016.

8. Z.M. Gill, M.J. Tierney, I.M. Pegg & N. Allan, 'Measured energy and water performance of an aspiring low energy/carbon affordable housing site in the UK', *Energy and Buildings*, vol. 43, no. 1, 2011, p 117–25.

9. BSI, *BS 8536-1: 2015 Briefing for design and construction. Code of practice for facilities management (Buildings infrastructure)*, London, British Standards Institution, 2015, pp i–iv,1–88.

10. R. Hay, F. Samuel, K.J. Watson & S. Bradbury, 'Post-occupancy evaluation in architecture: Experiences and perspectives from UK practice', *Building Research & Information*, 2017, pp 1–13.

11. T. Gerrish, et al., 'Analysis of basic building performance data for identification of performance issues', *Facilities*, vol. 35 nos 13–14, 2017, pp 801–17.

12. R. Hay, et al., 'Pathways to POE, Value of architects', University of Reading, RIBA, 2017.

13. J. Williams, B. Humphries & A. Tait, *Post Occupancy Evaluation and Building Performance Evaluation Primer,* London, RIBA, 2016.

14. G.T. Moore, 'Teaching design evaluation, with results from case studies of playgrounds, schools, and housing for the elderly', *Design Studies*, vol. 4, no. 2, 1983, pp 100–14.

15. A. Kwok, W. Grondzik & B. Hagland, 'Academic advocacy: Teaching outside the academy', *Conference Proceedings of the 27th International Conference on Passive and Low Energy Architecture*, Louvain, Belgium, 2011.

16. Ibid.

17. J.D. Quale, *Sustainable, Affordable, Prefab: The Ecomod Project*, Charlottesville, University of Virginia Press, 2012.

18. E. Rodriguez-Ubinas, S. Rodriguez, K. Voss & M.S. Todorovic, 'Energy efficiency evaluation of zero energy houses', *Energy & Buildings*, 83, 2014, pp 23–35.

19. R. Gupta, 'Leading by example: Post-occupancy evaluation studies of city council-owned non-domestic buildings in Oxford to assess the potential for reducing CO_2 emissions', *24th International Conference on Passive and Low Energy Architecture (PLEA)*, Singapore, 2007.

20. F. Stevenson, A. Roberts & S. Altomonte, 'Designs on the planet: A workshop series on architectural education and the challenges of climate change', *Proceedings of the 26th International Conference on Passive and Low Energy Architecture*, Quebec, 2009.

21. SCHOSA, SCHOSA Conference: *Beyond Building Performance: Architectural Research, Practice and Education*, London, 2015.

22. RIBA, *RIBA Procedures for Validation and Validation Criteria for UK and International Courses and Examinations in Architecture*, London, RIBA, 2014.

23. D. King, P. MacCombie & S. Arnold, *The Case for Centres of Excellence in Sustainable Design*, London, The Royal Academy of Engineering, 2012.

24. N. Young, et al., 'How do potential knowledge users evaluate new claims about a contested resource? Problems of power and politics in knowledge exchange and mobilization', *Journal of Environmental Management*, 184, 2016, pp 380–8.

25. C. Eaton, 'GHA monitoring programme 2011–13: Technical report. Results from phase 2: post-

occupation testing of a sample of sustainable new homes', London, Good Homes Alliance, 2014.

26. C. Gaze, 'Information paper: lessons from AIMC4 for cost-effective, low-energy, fabric-first housing: Part 5: As-built performance and Post-occupancy evaluation', Watford, Building Research Establishment, 2014.

27. This was called the 'Carbon Challenge', and I was employed to develop the BPE process with the Carbon Challenge team in 2009.

28. M. Baborska-Narożny, Magdalena, F. Stevenson & P. Chatterton, 'A social learning tool: Barriers and opportunities for collective occupant learning in low carbon housing', *Energy Procedia*, 62, 2014, pp 492–501.

29. F. Wade, M. Shipworth & R. Hitchings, 'How installers select and explain domestic heating controls', *Building Research & Information*, vol. 45, issue 4, 2017, pp 371–83.

Chapter 10

1. H. Shahrokhi, et al., 'A comparative analysis of different post-occupancy building assessment standards', Graduate Research, University of British Columbia, 2016.

2. EVO, 'Core concepts: International performance measurement and verification protocol', Washington DC, *Efficiency Valuation Organisation*, USA, 2016.

3. D.J. McElroy & J. Rosenow, 'Policy implications for the performance gap of low-carbon building technologies', *Building Research & Information*, 2018, pp 1–13.

4. R.J. Cole & M. Jose Valdebenito, 'The importation of building environmental certification

systems: International usages of BREEAM and LEED', *Building Research & Information*, vol. 41, no. 6, 2013.

5. Z. Gou & S.S-Y. Lau, 'Contextualizing green building rating systems: Case study of Hong Kong', *Habitat International*, 44, 2014, pp 282–9.

6. P. Rohdin, A. Molin & B. Moshfegh, 'Experiences from nine passive houses in Sweden: Indoor thermal environment and energy use', *Building and Environment*, 71, 2014, pp 176–85.

7. S. Peper & W. Feist, 'Die Energieeffizienz des Passivhaus-Standards: Messungen bestätigen die Erwartungen in der Praxis, *Passivhaus Institut*, 2015.

8. F. Stevenson & M. Baborska-Narożny, 'Housing performance evaluation: Challenges for international knowledge exchange', *Building Research & Information*, 2017, pp 1–12.

9. Ibid.

10. S. Villa, F. Garrefa, F. Stevenson & K. Bortoli, 'Co-production and resilience in a Brazilian Social Housing: the case of "Shopping Park" neighbourhood', *Passive Low Energy Architecture International Conference*, 2017.

11. D.C.C.K. Kowaltowski, et al., 'A critical analysis of research of a mass-housing programme', *Building Research & Information*, vol. 47, no. 6, 2019, pp 716–33.

12. S. Barbosa Villa, F. Garrefa, F. Stevenson, A. Ribeiro Souza, K. Carrer Ruman de Bortoli, J. Silva Arantes, P. Barcelos Vasconcelos & V. Araujo Campelo, [RESAPO_stage 1] Method of Analysis of the Resilience and Adaptability in Social Housing Developments through Post-occupancy Evaluation and Co-

production Research. Final report. FAUed/UFU. Federal University of Uberlandia, Brazil, 2017, pp 1–403.

13. K. Carrer Ruman de Bortoli. *Avaliando a Resiliência no Ambiente Construído: Adequação Climática e Ambiental Em Habitações de Interesse Social no Residencial Sucesso Brasil* (Uberlândia/MG), Dissertação (Mestrado em Arquitetura e Urbanismo), Universidade Federal de Uberlândia, Brazil, 2018, p 282.

14. Ibid.

15. T. Williamson, V. Soebarto & A. Radford, 'Comfort and energy use in five Australian award-winning houses: Regulated, measured and perceived', *Building Research & Information*, vol. 38, no. 5, 2010, p. 509–29.

16. Ibid.

17. T. Moore, et al., 'Dwelling performance and adaptive summer comfort in low-income Australian households', *Building Research & Information*, vol. 45, no. 4, 2017, pp 443–56.

18. R. Rovers, 'New energy retrofit concept: "Renovation trains" for mass housing', *Building Research & Information*, vol. 42, no. 6, 2014, p 757–67.

19. http://bpie.eu/publication/renovation-in-practice/

20. P. Li, T.M. Froese & G. Brager, 'Post-occupancy evaluation: State-of-the-art analysis and state-of-the-practice review', *Building and Environment*, 133, 2018, pp 187–202.

Chapter 11

1. R. Gupta, M. Gregg, S. Passmore & G. Stevens, 'Intent and outcomes from the Retrofit for the Future programme: Key lessons', *Building Research & Information*, vol. 43, no. 4, 2015, pp 435–51.

2. F. Stevenson, 'Embedding building performance

evaluation in UK architectural practice and beyond', *Building Research & Information*, vol. 47, no. 3, 2019, pp 305–17.

3. N. May & C. Sanders, *Moisture in Buildings: An Integrated Approach to Risk Assessment and Guidance*, London, BSI, 2016.

4. L. Gooding & M.S. Gul, 'Achieving growth within the UK's domestic energy efficiency retrofitting services sector: Practitioner experiences and strategies moving forward', *Energy Policy*, 105, 2017, pp 173–82.

5. R. Gupta & S. Chandiwala, 'Understanding occupants: Feedback techniques for large-scale low-carbon domestic refurbishments', *Building Research & Information*, vol. 38, no. 5, 2010, pp 530–48.

6. R. Gupta & M. Gregg, 'Do deep low carbon domestic retrofits actually work?', *Energy and Buildings*, 129, 2016, pp 330–43.

7. L. Vlasova & K. Gram-Hanssen, 'Incorporating inhabitants' everyday practices into domestic retrofits', *Building Research & Information*, vol. 42, no. 4, 2014, pp 512–24.

8. R. Gupta & S. Chandiwala, 2010.

9. R. Gupta & M. Gregg, 2016.

10. M. Sunikka-Blank & R. Galvin, 'Introducing the prebound effect: The gap between performance and actual energy consumption', *Building Research & Information*, vol. 40, no. 3, 2012, pp 260–73.

11. R. Gupta & M. Gregg, 2016.

12. P. Chatterton, *Low Impact Living: A Field Guide to Ecological, Affordable Community Building*, London, Routledge, 2015.

13. M. Baborska-Narożny & F. Stevenson, 'Service controls interfaces in housing: Usability and engagement tool development', *Building Research & Information*, vol. 47, no. 3, 2019, pp 290–304.

14. M. Baborska-Narożny, F. Stevenson & P. Chatterton, 'A social learning tool: Barriers and opportunities for collective occupant learning in low carbon housing', *Energy Procedia*, 62, 2014, pp 492–501.

15. M. Baborska-Narożny & F. Stevenson, 'Mechanical ventilation in housing: Understanding in-use issues', *Engineering Sustainability*, 170, 2017.

16. M. Baborska-Narożny, F. Stevenson & P. Chatterton, 'Temperature in housing: Stratification and contextual factors', *Engineering Sustainability*, 169, 2016, p 125–37.

17. M. Baborska-Narożny, F. Stevenson & P. Chatterton, 2014.

18. F. Stevenson, M. Baborska-Narożny & P. Chatterton, 'Resilience, redundancy and low-carbon living: Co- producing individual and community learning', *Building Research & Information*, vol. 44, no. 7, 2016, pp 789–803.

Chapter 12

1. R. Hay, F. Samuel, K.J. Watson & S. Bradbury, 'Post-occupancy evaluation in architecture: Experiences and perspectives from UK practice', *Building Research & Information*, vol. 46, no. 6, 2018, pp 698–710.

2. Ibid., pp 1–13.

3. R. Hay, et al., *Pathways to POE, Value of Architects*, University of Reading, RIBA, 2017.

4. Ibid.

5. R. Hay, F. Samuel, K.J. Watson & S. Bradbury, 2017.

6. P. Chatterton, *Low Impact Living: A Field Guide to Ecological, Affordable Community Building*, London, Routledge, 2015.

7. A. Moody, *Building Performance Evaluation: Industry Capability and Appetite*, London, Innovate UK, 2013.

8. F. Stevenson & H.B. Rijal, 'Developing occupancy feedback from a prototype to improve housing production', *Building Research & Information*, 38, 2010, pp 549–63.

9. 'CE298: Monitoring energy and carbon performance in new homes', London, Energy Savings Trust, 2008.

10. A. Moody, 2013.

11. T. Moore, Y. Strengers & C. Maller, 'Utilising mixed methods research to inform low-carbon social housing performance policy', *Urban Policy and Research*, 34, 2016, pp 240–55.

12. Z. Frances & F. Stevenson, 'Domestic photovoltaic systems: The governance of occupant use', *Building Research & Information*, 46, 2018, pp 23–41.

13. M. Baborska-Narożny, F. Stevenson & F.J. Ziyad, 'User learning and emerging practices in relation to innovative technologies: A case study of domestic photovoltaic systems in the UK', *Energy Research & Social Science*, 13, 2016, pp 24–37.

14. F. Stevenson & H.B. Rijal, 2010.

15. P. Burns & J. Coxon, *Boilers on Prescription Trial: Closing Report*, Sunderland, UK, Gentoo Group, 2016.

16. K.J. Watson, J. Evans, A. Karvonen & T. Whitley, 'Capturing the social value of buildings: The promise of Social Return on Investment (SROI)', *Building and Environment*, 103, 2016, pp 289–301.

17. Verco, *Building Performance Evaluation: Monitoring of an Estate Near Rotherham*, London, Innovate UK 2014.

18. T. Morton, *Earth Masonry: Design and Construction*

Guidelines, Berkshire, IHS
BRE Press, 2008.

19. R. Lowe, J. Wingfield, M.
    Bell & J. Bell, 'Evidence
    for heat losses via party
    wall cavities in masonry
    construction', *Building
    Services Engineering
    Research & Technology*, 28,
    2007, pp 161–81.

Chapter 13

1. T. Sharpe, 'Ethical issues
   in domestic building
   performance evaluation
   studies', *Building Research
   & Information*, vol. 47, no.
   3, 2019, pp 318–29.
2. A. Hope, et al., 'The role
   of compensatory beliefs in
   rationalizing environmentally
   detrimental behaviors',
   Environment and Behavior,
   vol. 50, no. 4, 2018, pp
   401–25.
3. B. Bordass & A. Leaman,
   'A new professionalism:
   remedy or fantasy?',
   *Building Research &
   Information*, vol. 41, no. 1,
   2013, pp 1–7.
4. F. Stevenson, M. Baborska-
   Narożny & P. Chatterton,
   'Resilience, redundancy
   and low-carbon living:
   Co-producing individual
   and community learning',
   *Building Research &
   Information*, vol. 44, no. 7,
   2016, pp 789–803.
5. J. Balvers, et al.,
   'Mechanical ventilation
   in recently built Dutch
   homes: Technical
   shortcomings, possibilities
   for improvement, perceived
   indoor environment and
   health effects', *Architectural
   Science Review*, vol. 55,
   no. 1, 2012, pp 4–14.
6. T. Sharpe, 2019.
7. A. Gruzd, et al., 'Exploring
   the efficacy of Facebook
   groups for collective
   occupant learning about
   using their homes',
   *American Behavioral
   Scientist*, vol. 61, no. 7,
   2017, pp 757–73.
8. E. Shove, 'What is wrong
   with energy efficiency?',
   *Building Research &

Information, 2017, pp 1–11.
9. I. Cooper, 'The socialization
   of building science: The
   emblematic journey of R.J.
   Cole', *Building Research &
   Information*, vol. 46, no. 5,
   2018, pp 463–8.
10. T. Sharpe, 2019.
11. A. Leaman, F. Stevenson
    & B. Bordass, 'Building
    evaluation: Practice
    and principles', *Building
    Research & Information*, vol.
    38, no. 5, 2010, pp 564–77.
12. G. McGill, et al., 'Meta-
    analysis of indoor
    temperatures in new-build
    housing', *Building Research
    & Information*, vol. 45, nos
    1–2, 2017, pp 19–39.
13. R. Gupta, M. Gregg, S.
    Passmore & G. Stevens,
    'Intent and outcomes from
    the Retrofit for the Future
    programme: Key lessons',
    *Building Research &
    Information*, vol. 43, no. 4,
    2015, pp 435–51.
14. P. Desai, *One Planet
    Communities: A Real-Life
    Guide to Sustainable Living*,
    Hoboken, NJ, Wiley, 2010.
15. R. Hay, et al., *Pathways to
    POE, Value of Architects*,
    University of Reading, RIBA,
    2017.
16. F. Stevenson, 'Embedding
    building performance
    evaluation in UK
    architectural practice and
    beyond', *Building Research
    & Information*, vol. 47, no.
    3, 2019, pp 305–17.
17. F. Stevenson & H.B. Rijal,
    'Developing occupancy
    feedback from a prototype
    to improve housing
    production', *Building
    Research & Information*, vol.
    38, no. 5, 2010, pp 549–63.

Chapter 14

1. F. Duffy, 'Forum linking
   theory back to practice',
   *Building Research &
   Information*, 36, 2008, pp
   655–8.
2. W.F.E. Preiser & J. Vischer,
   *Assessing Building
   Performance*, London,
   Elsevier Butterworth-
   Heinemann, 2005.

3. R. Rovers, 'New
   energy retrofit concept:
   "Renovation trains" for
   mass housing', *Building
   Research & Information*, 42,
   2014, pp 757–67.
4. A. Leaman, F. Stevenson
   & B. Bordass, 'Building
   evaluation: Practice
   and principles', *Building
   Research & Information*, 38,
   2010, pp 564–77.
5. R. Hay, et al., *Pathways to
   POE, Value of Architects*,
   University of Reading, RIBA,
   2017.

Chapter 15

1. R. Cohen, et al., 'How the
   commitment to disclose
   in-use performance can
   transform energy outcomes
   for new buildings', *Building
   Services Engineering
   Research & Technology*, 38,
   2017, pp 711–27.
2. F. Stevenson, 'Embedding
   building performance
   evaluation in UK
   architectural practice and
   beyond', *Building Research
   & Information*, 47, 2019, pp
   305–17.
3. J. Williams, B. Humphries
   & A. Tait, *Post Occupancy
   Evaluation and Building
   Performance Evaluation
   Primer*, London, RIBA,
   2016.
4. BEIS, *The Clean Growth
   Strategy: Leading the Way
   to a Low Carbon Future*,
   London, Department for
   Business, Energy and
   Industrial Strategy, 2017,
   p 81.
5. E. Shove, 'Beyond the
   ABC: Climate Change
   Policy and Theories
   of Social Change',
   *Environment and Planning
   A*, 42, 2010, pp 1273–85.
6. T. Hargreaves, C. Wilson
   & R. Hauxwell-Baldwin,
   'Learning to live in a smart
   home', *Building Research
   & Information*, 46, 2018, pp
   127–39.
7. A. Leaman & B. Bordass,
   'Productivity in buildings:
   The "killer" variables',
   *Building Research &

Information, 27, 1999, pp 4–19.

8. D. Teli, et al., 'Fuel poverty-induced "prebound effect" in achieving the anticipated carbon savings from social housing retrofit', *Building Services Engineering Research & Technology*, 37, 2016, pp 176–93.

9. C. Grandclément, A. Karvonen & S. Guy, 'Negotiating comfort in low energy housing: The politics of intermediation', *Energy Policy*, 84, 2015, pp 213–22.

10. F. Stevenson, M. Baborska-Narożny & P. Chatterton, 'Resilience, redundancy and low-carbon living: Co-producing individual and community learning', *Building Research & Information*, 44, 2016, pp 789–803.

11. N. Schoon & E. Auckland, *Measuring Up: How the UK is Performing on the UN Sustainable Development Goals*, UK Stakeholders for Sustainable Development, 2018.

12. CCC, *Reducing UK Emissions 2018 Progress Report to Parliament*, London, Committee on Climate Change, 2018b, p 95.

13. CCC, *An Independent Assessment of the UK's Clean Growth Strategy: From Ambition to Action*, London, Committee on Climate Change, 2018a.

14. T. Hargreaves, C. Wilson & R. Hauxwell-Baldwin, 'Learning to live in a smart home', *Building Research & Information*, 46, 2018, pp 127–39.

15. R.J. Cole, J. Robinson, Z. Brown & M. O'Shea, 'Re-contextualizing the notion of comfort', *Building Research & Information*, 36, 2008, pp 323–36.

16. G. Sousa, B. Jones, P. Mirzaei & D. Robinson, 'A review and critique of UK housing stock energy models, modelling approaches and data sources', *Energy and Buildings*, 151, 2017, pp 66.

17. M. Shipworth, et al., 'Central heating thermostat settings and timing: Building demographics', *Building Research & Information*, 38, 2010, pp 50–69.

18. H. Birgisdottir, et al., 'IEA EBC annex 57 "evaluation of embodied energy and CO_2eq for building construction"', *Energy and Buildings*, 154, 2017, pp 72–80.

19. T. Hargreaves, C. Wilson & R. Hauxwell-Baldwin, 2018.

20. CCC, 2018b.

21. Ibid, p 21.

22. Ibid, p112.

23. BEIS, 2017.

24. CCC, 2018a.

25. Ibid.

26. J. Ross, 'Cavity wall insulation inspection final report', *Northern Ireland Housing Executive*, 2014.

27. E.Shove, 'What is wrong with energy efficiency?', Building Research and Information, 2017, pp 1-11.

Primer

1. https://www.usgbc.org/discoverleed/certification/homes/

2. https://www.usgbc.org/discoverleed/certification/homes/)

3. https://living-future.org/affordable-housing/

4. https://v2.wellcertified.com/landing

5. http://www.homequalitymark.com/filelibrary/Technical%20consultation/HQM-ONE-Draft-Manual--Consultation-.pdf

6. https://www.bre.co.uk/filelibrary/SAP/2012/SAP-2012_9-92.pdf

7. http://www.passivhaus.org.uk/

8. http://www.irtraining.eu/en/

9. https://www.busmethodology.org.uk/partner.html

10. https://www.leedsbeckett.ac.uk/as/cebe/projects/iea_annex58/whole_house_heat_loss_test_method(coheating).pdf

11. https://shop.bsigroup.com/Browse-by-Sector/Building--Construction/Whitepaper-Moisture-in-buildings/

12. https://www.sciencedirect.com/science/article/pii/S0160412016309989

13. https://sites.google.com/view/bpe-poland/bpe-poland/bupesa-project/tools/usability-survey

14. S. Pink, et al., *Making Homes: Ethnography and Design*, London, UK Bloomsbury Academic, 2017.

15. R. Barbour, *Doing Focus Groups*, London: Sage Publications, 2007.

16. C. Robson, *Real World Research: A Resource for Users of Social Research Methods in Applied Settings*, Chichester, Wiley, 2011, p 190.

Index

Note: Page numbers in *italics* refer to text that includes maps, plans or illustrations

Image credits

Cover image © Barratt Homes &HTA; Design LLP pvii © Louis Hellman; pviii © Fionn Stevenson; p9 © Fionn Stevenson; p5 Painting 'The Doctor' by Luke Fildes, 1924; p6 Diagram redrawn from Watt, D, 'Building Pathology', 2007 p 91; p10 'Redrawn Joseph Bradley by from David Lithgow https://creativecommons.org/licenses/by/3.0/deed.en; p12 *top* © Fionn Stevenson, *bottom* Diagram drawn William Capps; p13 © Gokay Deveci; p16 © US National Oceanic and Atmospheric Administration satellite; p18 Extract redrawn by Joseph Bradley from http://bit.ly/1VoGGpT; p20 via Creative Commons; p23 via Creative Commons; p 24 Photographed by Adrian Pingstone; p29 Diagram drawn by Joseph Bradley; p30 © Fionn Stevenson; p32 Diagram drawn by Joseph Bradley; p34 Diagram drawn by Joseph Bradley; p38 Image Creative Partnerships Ltd and © Barratt Homes &HTA Design LLP; p41 Diagram drawn by William Capps; p43 Diagram drawn by Joseph Bradley; p45 *both* © Magda Baborska Narozny; p48 Diagram drawn by William Capps; p 52 Diagram drawn by William Capps; p 54 Diagram drawn by William Capps; p56 © Madga Baborska Narozny; p59 © Madga Baborska Narozny; p62 © Madga Baborska Narozny; p65 Adapted by Chris Morgan from Stewart Brand's book *How Buildings Learn*; p66 ©Gokai Deveci; p69 © Fionn Stevenson; p70 © William Bordass, Adrian Leaman and Joanne Eley; p71 © Chris Kendrick; p81 *left* © Andrew Elliot, *right* © Dario Sabljak/Shutterstock.com; p82 © Magda Baborska Narozny; p84 © Isabel Carmona redrawn by Joseph Bradley; p86 © Chris Kendrick; p88 © Chris Kendrick; p93 © Stewart Milne Group; p94© Magda Baborska Narozny; p95 chart by Isabel Carmona; p99 © Luke Mills, EcoArc architects; p103 *top* Drawn by Joseph Bradley, *bottom* Drawn by William Capps; p104 © Darren Carter/ Morgan Sindall; p105 © Fionn Stevenson; p108 © Chris Kendrick; p117 © Ana Carolina de Oliveira Stefani; p118 © Fionn Stevenson; p119 © Karen Carrer Ruman de Bortoli; p120 Photographed by Ryan Wilson; p121 Drawn by William Capps; p123 © Energiesprong, drawn by William Capps; p124 Drawn by William Capps; p127 English Housing Survey 2017-2018 Redrawn by William Capps; p129 © Fionn Stevenson; p131 *both* © Rajat Gupta; p133 © Magda Baborska Narozny; p134–5 © Magda Baborska Narozny; p136 © Magda Baborska Narozny; p142 Redrawn by William Capps; p149 © Magda Baborska Narozny; p150 *top* © Stewart Milne Group, *bottom* © Fionn Stevenson; p151 Drawn by Joseph Bradley; p159 © Chris Kendrick; p160 *both* Dario Sabljak/Shutterstock.com; p167 Drawn by William Capps; p168 JD Lasica via Creative Commons; p169 via Creative Commons; p172 Photographed by Luke Mills, EcoArc Architects; p178 © Fionn Stevenson; p179 Monkey Business Images/Shutterstock.com; p182 Zhu Difeng/Shutterstock.com; p189 *both* © Fionn Stevenson; p196-7 © Magda Baborska Narozny; p199 © Fionn Stevenson; p204 chart by Isabel Carmona;